T0302257

Big Data Analysis for Green Computing

Green Engineering and Technology: Concepts and Applications

Series Editor:
Brujo Kishore Mishra
GIET University, India and Raghvendra Kumar, LNCT College, India

Environment is an important issue these days for the whole world. Different strategies and technologies are used to save the environment. Technology is the application of knowledge to practical requirements. Green technologies encompass various aspects of technology which help us reduce the human impact on the environment and create ways of sustainable development. Social equability, this book series will enlighten the green technology in different ways, aspects, and methods. This technology helps people to understand the use of different resources to fulfill needs and demands. Some points will be discussed as the combination of involuntary approaches, government incentives, and a comprehensive regulatory framework will encourage the diffusion of green technology, least developed countries and developing states of small island requires unique support and measure to promote the green technologies.

Convergence of Blockchain Technology and E-Business
Concepts, Applications, and Case Studies
*Edited by D. Sumathi, T. Poongodi, Bansal Himani,
Balamurugan Balusamy, and Firoz Khan K P*

Handbook of Sustainable Development Through Green
Engineering and Technology
*Edited by Vikram Bali, Rajni Mohana, Ahmed Elngar,
Sunil Kumar Chawla, and Gurpreet Singh*

Integrating Deep Learning Algorithms to Overcome Challenges in
Big Data Analytics
Edited by R. Sujatha, S. L. Aarthy, and R. Vettri Selvan

Big Data Analysis for Green Computing
Concepts and Applications
*Edited by Rohit Sharma, Dilip Kumar Sharma,
Dhowmya Bhatt, and Binh Thai Pham*

For more information about this series, please visit: https://www.routledge.com/ Green-Engineering-and-Technology-Concepts-and-Applications/book-series/ CRCGETCA

Big Data Analysis for
Green Computing
Concepts and Applications

Edited by
Rohit Sharma, Dilip Kumar Sharma,
Dhowmya Bhatt, and Binh Thai Pham

CRC Press
Taylor & Francis Group
Boca Raton London New York

CRC Press is an imprint of the
Taylor & Francis Group, an **informa** business

First edition published 2022
by CRC Press
6000 Broken Sound Parkway NW, Suite 300, Boca Raton, FL 33487-2742

and by CRC Press
2 Park Square, Milton Park, Abingdon, Oxon, OX14 4RN

© 2022 Taylor & Francis Group, LLC

CRC Press is an imprint of Taylor & Francis Group, LLC

Library of Congress Cataloging-in-Publication Data
Names: Sharma, Rohit, editor. | Śarmā, Dilīpa, editor. | Bhatt, Dhowmya, editor. |
Pham, Binh Thai, editor.
Title: Big data analysis for green computing : concepts and applications / edited
by Rohit Sharma, Dilip Kumar Sharma, Dhowmya Bhatt and Binh Thai Pham.
Description: First edition. | Boca Raton : CRC Press, [2022] |
Series: Green engineering and technology : concepts and applications |
Includes bibliographical references and index.
Identifiers: LCCN 2021026747 (print) | LCCN 2021026748 (ebook) |
ISBN 9780367442309 (hbk) | ISBN 9781032053066 (pbk) | ISBN 9781003032328 (ebk)
Subjects: LCSH: Big data.
Classification: LCC QA76.9.B45 B49945 2022 (print) |
LCC QA76.9.B45 (ebook) | DDC 005.7—dc23
LC record available at https://lccn.loc.gov/2021026747
LC ebook record available at https://lccn.loc.gov/2021026748

ISBN: 978-0-367-44230-9 (hbk)
ISBN: 978-1-032-05306-6 (pbk)
ISBN: 978-1-003-03232-8 (ebk)

DOI: 10.1201/9781003032328

Typeset in Times
by codeMantra

Contents

Preface

The main objective of this book publication is to explore the concepts of big data in green computing along with the recent research and development. It also includes various real-time applications and case studies in the field of engineering and technologies used. As populations grow and resources become scarcer, the efficient usage of these limited goods becomes more important.

Chapter 1 has a primary focus on different MCDM techniques for mitigating pollution based on fuzzy sets, which further generate feasible decision alternatives. The other is to serve as a platform for understanding how such methods can be formulated from computational thinking that can help in making informed decisions of any complex and less structured problem. Using such concepts, one can transform knowledge-based intuitions to mathematics and make strategic decisions.

Chapter 2 introduces Internet of Medical Things with respect to the multiple challenges like privacy, security, and low-power operation faced while incorporating IoT with healthcare sector. For emergency treatment, a cloud-based healthcare framework provides a platform to support patients over internet communication by medical experts. Privacy and security to the medical records are on higher priority due to the very sensitive nature of the content.

Chapter 3 demonstrates the relationship of big data with cloud computing by discussing the rise of each and the dependency of one over the other.

Chapter 4 introduces the role of the measurement in the decision making, contrasting the perspective of the batch and online data processing. Also, differentiation among the concepts such as the data, information, and knowledge is introduced considering the decision-making process.

Chapter 5 displays the fundamental needs of vitality protection and security in the cloud investigations. Distributed of computing is a foundation for running endeavor and web applications in financially savvy way. Be that as it may, the developing needs of cloud have expanded the vitality utilization of server farms, which has become a significant issue.

Chapter 6 speaks of a cleverly minimal effort home mechanization process that is built using IoT. With the aid of this framework, every home apparatus and electronic gadgets can be effortlessly controlled and shown through a web. Using this structure, the metering strategy for a home can likewise be guided.

Chapter 7 is intended to introduce big data and machine learning concepts to the reader while highlighting the importance of including green practices to make these technologies more effective.

Chapter 8 presents a complete perspective about blockchain technology, security risks, and challenges that prevail due to the vulnerability of being a popular technology. The authors summarize the empowering technologies for being scalable, privacy-free, and honest mining blockchain systems.

Chapter 9 contributes to various factors responsible for mental health problems, thereby leading to mental illness. These factors include stress, anxiety, depression, obsession, and obsessive disorder. Also, the chapter focuses on the working on the need for more effective mental healthcare systems.

Chapter 10 explains the usage of blockchain technology with various industry aspects along with Internet of Things (IoT), which is a very popular and auspicious technology used nowadays. There still exist some security problems and challenges resulting while implementing blockchain to industry 4.0 which opens a forum for future research direction.

MATLAB® is a registered trademark of The MathWorks, Inc. For product information, please contact:
The MathWorks, Inc.
3 Apple Hill Drive
Natick, MA 01760-2098 USA
Tel: 508-647-7000
Fax: 508-647-7001
E-mail: info@mathworks.com
Web: www.mathworks.com

Editors

Rohit Sharma is currently an Assistant Professor in the Department of Electronics and Communication Engineering, SRM Institute of Science and Technology, Delhi NCR Campus Ghaziabad, India. He is an active member of ISTE, IEEE, ICS, IAENG, and IACSIT. He is an editorial board member and reviewer of more than 12 international journals and conferences, including the topmost journal *IEEE Access and IEEE Internet of Things Journal.* He serves as a Book Editor for seven different titles to be published by CRC Press, Taylor & Francis Group, USA, Apple Academic Press, Springer, etc. He has received the Young Researcher Award in "2nd Global Outreach Research and Education Summit & Awards 2019" hosted by Global Outreach Research & Education Association (GOREA). He is serving as a Guest Editor in SCI journal of Elsevier, CEE, and Springer WPC. He has actively been an organizing end of various reputed international conferences. He is serving as an Editor and Organizing Chair to the Third Springer International Conference on Microelectronics and Telecommunication (2019) and has served as the Editor and Organizing Chair to the Second IEEE International Conference on Microelectronics and Telecommunication (2018), Editor and Organizing Chair to IEEE International Conference on Microelectronics and Telecommunication (ICMETE-2016) held in India, Technical Committee member in "CSMA2017 in Wuhan, Hubei, China", "EEWC 2017 in Tianjin, China" IWMSE2017 in Guangzhou, Guangdong, China", "ICG2016 in Guangzhou, Guangdong, China" "ICCEIS2016 in Dalian Liaoning Province, China".

Dilip Kumar Sharma has B.E. (CSE), M.Tech. (CSE), and Ph.D. in Computer Engineering. He is a Senior Member of IEEE, USA, ACM, USA, and CSI India and Fellow of IETE and IE (India). He has delivered/chaired more than 75 invited talks/guest lectures and chaired the technical sessions at various institutes/conferences. He has attended 32 short-term courses/workshops/seminars organized by various esteemed originations and edited three books and worked as a Guest Editor of international journals of repute. He has organized more than 12 IEEE/CSI International/National Conferences and Workshops with the capacity of General Chair, Co-General Chair, Convener and co-convener, etc. He has published more than 120 research papers in international journals/conferences of repute indexed in SCI, Scopus, and DBLP databases and participated in 50 international/national conferences. He has guided three PhD theses. He has received a Conferred Significant Contribution Award from Computer Society of India in 2012, 2013, 2014, and 2017. He is presently Associate Dean (Academic Collaboration) of GLA University and a Professor in the Department of Computer Engineering and Applications, GLA University Mathura India.

Dhowmya Bhatt is currently an Associate Professor in the Department of Information Technology, SRM IST, Delhi NCR Campus Ghaziabad, India. The research focus of her doctoral degree is network security. Dr Dhowmya has published many research papers in international journals of repute. In 2012, she organized National Conference on Recent Development in Computational and Information Technology NCRDCIT'12, which was funded by DRDO. She has been the secretary of International Conference on Recent Development in Computational and Information Technology, ICRDCIT in 2014. She has organized many workshops, seminars, and expert lectures in association with TCS, DRDO, and Cisco. She has been the Convener of many technical and cultural programs in SRM IST Delhi NCR campus. She has served as a session chair in conferences. Dr. Dhowmya is in the advisory committee of ICFCCT 19.

Binh Thai Pham is a senior researcher, lecturer, and Head of the Department of Geotechnical Engineering at University of Transport Technology, Hanoi, Vietnam. He received a PhD degree in Civil Engineering from the Gujarat Technological University, Gujarat, India. His research focuses on geotechnical engineering problems, machine learning, artificial intelligence, natural hazard modeling and assessment, geographic information system, remote sensing, and construction materials. Dr Binh Thai Pham has published about 80 scientific publications mainly on peer-reviewed international journals indexed SCI, SCIE, and SCOPUS. He is also a scientific reviewer of more than 30 ISI journals.

Contributors

Pooja Agarwal
Computer Science and Engineering
PES University
Bengaluru, India

Puneet Kumar Aggarwal
Department of CSE
ABES
Ghaziabad, India

Arti Arya
Head, MCA Department
PES University
Bengaluru, India

R. Ganesh Babu
Department of Electronics and
    Communication Engineering
SRM TRP Engineering College
Tiruchirappalli, India

Bharat Bhushan
Computer Science and Engineering
School of Engineering and Technology,
    Sharda University
Greater Noida, India

J. Bino
St. Joseph's Institute of Technology
Chennai, India

Sneha Chaudhary
Computer Science
JIMS
Delhi, India

Charu Chhabra
Computer Science and Engineering
RDIAS
Delhi, India

M. J. Diván
Economy School
National University of La Pampa,
    Santa Rosa, Argentina

K. Elangovan
Siddharth Institute of Engineering and
    Technology
Puttur, India

Kashish Garg
Computer Science & Engineering
Chandigarh University
Ajitgarh, India

M. Gracy
Department of Computer Applications
CSH, SRMIST, KTR
Chengalpattu, India

Meenu Gupta
Computer Science & Engineering,
Chandigarh University
Ajitgarh, India

Megha Gupta
Department of CSE
IMS
Ghaziabad, India

Himanshu

Rachna Jain
Computer Science & Engineering
Bharati Vidyapeeth's College of
    Engineering
Navi Mumbai, India

P. Jayachandran
Department of Computer Science
Thiruthangal Nadar College
Chennai, India

B. Rebecca Jeyavadhanam
Associate Professor
Computer Applications
Department of Computer Applications,
    Kattankulathur Campus, SRM
    Institute of Science and Technology,
    Chennai (formerly known as SRM
    University).

Ila Kaushik
Krishna Institute of Engineering &
    Technology
Ghaziabad, India

S. Kalimuthu Kumar
Department of Biomedical Engineering
Kalasalingam Academy of Research
    and Education
Krishnankoil, India

Ayasha Malik
Computer Science and Engineering
Noida Institute of Engineering and
    technology
Greater Noida, India

G. Manikandan
Electronics and Communication
    Engineering
Dr. M.G.R Educational and Research
    Institute
Chennai, India

Sudhanshu Maurya
School of Computing
Graphic Era Hill University
Dehradun, India

D. Antony Joseph Rajan
Research Scholar
SCSVMV University
Chennai, India

V. V. Ramalingam
Department of Computer Science and
    Engineering
SRMIST, KTR
Chengalpattu, India

Rishu Rana
HMR Institute of Technology &
    Management
Delhi, India,

M. Sánchez-Reynoso
Economy School
National University of La Pampa
Santa Rosa, Argentina

Anupam Sharma
Department of CSE
HMR Institute of Technology and
    Management
New Delhi, India

Nikhil Sharma
Department of CSE
Delhi Technological University
Delhi, India

Mandeep Singh
Department of CSE
HMR Institute of Technology and
    Management
New Delhi, India

Saurabh Singhal
Department of Computer Engineering
    and Applications
GLA University, Mathura

Monica Sneha
PES University
Bengaluru, India

Namrata Sukhija
Department of CSE
HMR Institute of Technology and
    Management
New Delhi, India

Christopher Xavier
Barry-Wehmiller Design Group
Chennai, India

1 Multi-Criteria and Fuzzy-Based Decision Making

Applications in Environment Pollution Control for Sustainable Development

Meenu Gupta and Kashish Garg
Chandigarh University

Rachna Jain
Bharati Vidyapeeth's College of Engineering

CONTENTS

DOI: 10.1201/9781003032328-1

1

1.1 INTRODUCTION TO FUZZY

Lotfi A. Zadeh introduced fuzzy logic in the year 1965. Fuzzy logic is capable of dealing with imprecise and incomplete data. It uses linguistic variables to deal with inaccurate data in a precise manner. It has been observed that fuzzy logic can be used in the development of various applications such as smart systems for decision making, determination, identification, pattern identification and recognition, optimization, operation and control [1].

Fuzzy logic contains four main processing frameworks [2] as shown in Figure 1.1:

- Fuzzification
- De-fuzzification
- Knowledge base
- Inference engines

The fuzzification process converts the crisp sets into fuzzy sets. For a graphical representation of fuzzy sets, various membership functions are used for the purpose. Knowledge base consists of if-then verbalizations that are provided by the experts [1]. The inference engine will stimulate the human prospects by taking fuzzy inference as input the if-then rules that anteriorly set in knowledge base framework. De-fuzzification, with the help of the inference engine, changes the fuzzy sets into a value (output).

1.2 INTRODUCTION TO MCDM

Multi-criteria decision making (MCDM) plays a vital role in evaluating multiple criteria involved in decision making. It comes handy when one needs to select the best and optimum criteria from various conflicting approaches. These criteria can be chosen by providing weights to the standards. To evaluate the multiple criteria's essential aspect, the structure of the problem needs to be considered. Decision making is a process in which the alternatives are chosen based on judgments and pearls of wisdom of decision-makers. Decisions taken collectively (also known as group MCDM) are proved to be often impartial, unbiased, and productive than the decisions made individually [3]. In collective decision making, the decisions made by all the individuals are considered and combined to solve the given problem. The most crucial part of the decision-making approach is identifying and examining all the

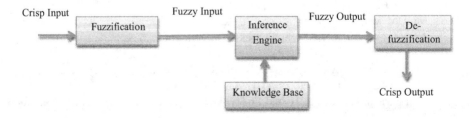

FIGURE 1.1   Working structure of fuzzy logic system.

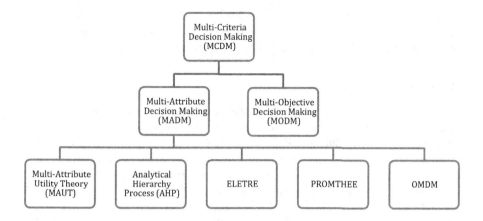

FIGURE 1.2 Classifications of MCDM methods [3].

criteria according to the assessment criteria. The alternatives must be ranked to gain knowledge about the most suitable option and to evaluate the respective priority of each option. Figure 1.2 shows the classification of MCDM.

MCDM refers to deciding the presence of multiple and incompatible criterias. A multi-criteria decision problem can consist of either numerous attributes, objectives or both. MCDM is classified into two parts, as shown in Figure 1.2: the first is MODM (multi-objective decision making) [4], and the second is MADM (multi-attribute decision making). MCDM has many applications, such as water resources planning, risk prediction, job evaluation, air pollution forecasting and business failure prediction [5].

1.2.1 MADM

MADM is used to solve the selections-related problem which is collected from a finite number of alternatives. This method proceeds with attribute information to arrive at a particular choice.

1.2.2 MODM

MODM is a special kind of problem which includes the design of various choices or alternatives available which helps to optimize and amplify the different objectives of the decision-maker. Consider an example of making a development plan for the government of a developing country [6]. The government would have various purposes while composing the acceptable policy such as maximizing the national welfare and minimizing the dependence on peregrine avail to reduce unemployment rate [3].

Mathematical representation of MODM problems is as follows [4]:

$$\text{Max}\left[f_1(x), f_2(x), \ldots, f_k(x)\right] \tag{1.1}$$

$$: g_i(x) \le 0, \, j = 1, \ldots, m \tag{1.2}$$

where x is an n-dimensional decision variable vector. The problem consists of n decision variables, m constraints and k objectives. Any or all of the functions may be nonlinear. These problems are also known as vector maximization problems.

1.2.3 NEEDS OF FUZZY MCDM

Fuzzy logic can give an approximate value even if the information is incomplete. In the case of complex problems, traditional methods related to non-fuzzy approach usually rely on mathematical estimates and calculations. Linearization of nonlinear problems is one of the examples which often gives bad functioning and is not very economical. Therefore, fuzzy systems surpass traditional MCDM techniques [7]. Decision-making problems are usually vague and uncertain in nature. Conventional MCDM methods cannot deal with this uncertainty. Therefore, the fuzzy set theory is applied to handle the possibility. Fuzzy MCDM techniques are decision analysis techniques merged with fuzzy techniques. Fuzzy techniques are commonly used with MCDM to deal with uncertainty and increase the accuracy of decision making [8]. Fuzzy MCDM is widely used in applications like energy, environment, source management, supplier selection and planning.

1.3 LITERATURE SURVEY

Different experts have different perceptions about the acceptable limits and the insufficiency in the parametric data of various air pollutants. As a result, it is seen that there are some built-in deficiencies while modeling out the perceptions of different experts. These mixed reviews can lead to various problems, so keeping that in mind, many researchers have started in building some applications using soft computing techniques which helps in the estimation of air quality indices [9]. Fuzzy logic plays an essential role in the process of converting the expert's experiments into mathematical languages, which indirectly helps in magnifying the usage of the model [10,11]. Many researchers have suggested predicting air quality indexes and monitoring air emissions using fuzzy logic. Many MCDM techniques have been used to study the impact of air pollution on socioeconomic development in various regions [12]. Some of them are listed below. In Ref. [13], the authors used MCDM as a technique to improve air quality in communities. They implemented Delphi method, analytical hierarchy process and fuzzy logic theory for reducing air pollution in urban areas.

   The authors in [14] discussed the problem to surveying the impact of air pollution as an MCDM problem. They used a novel MCDM technique that is TOPSIS (Technique for Order of Preference by Similarity to Ideal Solution) to understand the importance of every pollutant that is contributing to air pollution. Their method outperforms the traditional TOPSIS methods involving Bayesian regularization and the back-propagation (BP) neural network to optimize the weight (in the training process). They named the novel TOPSIS approach as smart MCDM technique. In contrast to conventional TOPSIS, here the entropy method is used to calculate the initial weights. In conventional TOPSIS, the masses were obtained from an expert's perspective [15]. The model was integrated with the Bayesian regularization and BP neural network architecture to boost the weights in the training loop [16]. In Ref. [9], the authors monitored the concentrations of SO_2 gas and PM_{10} for 4 years and used PROMETHEE/GAIA multi-criteria techniques for ranking pollution prone zones. In Ref. [17], the authors used MCDM tool for selecting the optimum machine for controlling air pollution and choosing the appropriate technology for air pollution filtering. TOPSIS method is used for selecting the optimum tool and strategy [18]. The primary purpose of this work was to opt an optimum control tool to reduce the cost and increase maintainability, efficiency and many other factors. In this chapter, every control criteria were provided with different weights for each air pollutant.

In Ref. [19], the authors proposed a method for determining air quality index using fuzzy logic. They named it as a fuzzy-based air quality index, and it was made using the fuzzy logic. Pollutants were given different weight factors according to their priorities. The priorities were assigned using trapezoidal membership functions, and the final index was computed using 72 inference rules. A case study was carried out to analyze the performance of the index, and according to the results, the author considered the method to be valid, definitive and useful. The local authorities can pay attention to assess air quality and building schemes to manage alarming situations. In Ref. [10], the authors discussed uncertainty in the multi-variate data of pollution and the cognitive change in outlining the pollutants by the domain experts in linguistic terms categorized as poor, good, very good, etc. She suggested that fuzzy logic can be used to model these two types of uncertainties. Air quality can be easily expressed in terms of linguistic variables having some certainty degree attached to them. In this work, the experts knowing air pollution described the air quality index in linguistic terms when the information and knowledge about pollutants were provided to them. The multi-variate pollution data were collected, whereas probability distribution was identified and fitted for determining the mean and variance. The probability distribution is converted into possibility distribution using Convex Normalised Fuzzy Number (CNFN). Fuzzy sets and numbers handled uncertainty in the beliefs of experts. Parameters determining air quality were given linguistic description [20].

Based on the experts' knowledge of the fuzzy domain, sets and numbers were generated to shape zones of trust for words such as very poor, poor, fair, good and very good [21]. The fuzzy rule-based scheme was executed and applied after involving four air quality experts. It was observed that specific values were attached to linguistic terms using the fuzzy logic model. This method was used to define air quality. This method was found more reliable than conventional methods where the numeric value is assigned to the air quality index and then classified in linguistic terms. Air quality index was predicted for Pune city using a fuzzy rule-based system, which proved to be more reliable than conventional methods [22]. Five different stages of pollution level – dangerous, bad, regular, good and excellent – were described by various authors, using the inference system of fuzzy for parameter classification along with reasoning process and while integrating it with an air quality index. The fuzzy inference system, which was used earlier, was also used in predicting the air quality index by building autoregressive models. These models were used to discover air quality concentrations [23].

Further, in Ref. [24], the authors have discovered an inverse association between environmental risk and income indicator in the city of Paranaguá, Paraná (Brazil). Researchers there used the fuzzy logic to predict environmental risk and confirmed using some of the income indicators. They claimed that the information provided by them provides a useful resource for more environmental risk assessments that may be used in the formulation of different public policies. According to the authors [25], air pollution is the most prominent climate risk, and many low socioeconomic situations exist in various regions solely because of this risk. The primary cause of air pollution is mainly the large ship engines, diesel ships and trains. Fuzzy logic proved to be a vital tool for monitoring air pollution in industries and in evaluating air pollution risk caused by industries. In Ref. [26], the authors considered fuzzy logic for grading the air pollutant risk, hazardous industries and high-threat areas by a constant value between zero and one [27].

Further, in Ref. [28], they used fuzzy inference system and fuzzy c-means clustering to analyze the air quality of Tehran in Iran. Fuzzy logic is also used with neural networks for the prediction of air pollution. In Ref. [29], they used a hybrid of artificial neural networks and fuzzy inference system for air quality forecasting. An artificial neural network learns the pattern by adjusting weights and interconnections between the layers, and a fuzzy inference system helped in incorporating the uncertainty. They used a three-layered neural fuzzy architecture with a BP learning algorithm.

Much research has been done in the prediction of air pollution using fuzzy and fuzzy hybrid models. Table 1.1 demonstrates the use of fuzzy logic and fuzzy hybrid systems in air pollution forecasting and air quality prediction [30].

TABLE 1.1
Air Quality Prediction and Air Pollution Forecasting Using Fuzzy Models

Models	Methodology
Fuzzy numbers, distance Function [1]	The fuzzy logic for air pollution forecasting was calculated using distance functions then predicted data generated fuzzy numbers. Actual data and predicted data were compared using distance function to calculate the grade of membership.
Hierarchical fuzzy inference system [31]	Hierarchical fuzzy inference systems are used for air pollution assessment. This system was built to assess the quality of air pollution.
Fuzzy logic algorithm [32]	A fuzzy logic algorithm is used for modeling and evaluating air quality. The fuzzy logic algorithm was used to predict the concentrations of air pollutants.
Neuro-fuzzy approach [33]	NO_2 pollutants addressed to air quality dispersion model over Delhi, India, were forecasted using neuro-fuzzy approach. Concentration of NO_2 was forecasted with the help of hybrid of neural networks and fuzzy logic which proved to be a better forecasting model than MLR and ANN models.

1.4   FUZZY CONTROL SYSTEM

In the fuzzy control system, different terminologies used in the fuzzy control system are fuzzification, rule assessment and de-fuzzification. Fuzzification used membership function to describe any situation graphically. This process can be considered as a process of transforming a crisp set to a fuzzy set or a fuzzy set to a fuzzier set. During the process, the accurate and crisp input gets translated into linguistic variables. Rule assessment is an application of fuzzy rules. Further, de-fuzzification is a process of obtaining crisp or actual results. It is the method of converting a fuzzy set into a crisp set (or converting a fuzzy member to a crisp member) [34].

1.4.1   Fuzzy Classification

Fuzzy classification can be defined as a process of making clusters of elements into a fuzzy set. The components used are those whose membership function is determined by the truth value of a fuzzy propositional function. Like the fuzzy set protracts the classical sets in a function, the fuzzy classification is also a native extension of the conventional classification. During the crisp classification, each object is assigned to exactly one class. This class describes the membership degree of the object, which is either 0 or 1. This makes all the objects mutually exclusive. In the fuzzy classification, an object belongs to many classes simultaneously, and each has a membership degree. This degree expresses as to what scope a particular object belongs to the different classes [27].

1.4.2   De-Fuzzification

De-fuzzification is the method of generating a computable outcome in crisp logic, where the fuzzy sets and corresponding membership degrees are already given. It is a process that is used to map a fuzzy set to a crisp set. This includes a number of rules that turn a large number of variables into a fuzzy result, which ensures that the outcome is represented in terms of membership in the fuzzy sets. For example, rules designed to determine how much pressure to apply could result in "Decrease Pressure (15%), Maintain Pressure (34%) and Increase Pressure (72%)." The method of de-fuzzification also includes the analysis of the membership degrees of the fuzzy sets into a particular judgment or actual value [21].

1.5   MATHEMATICAL PROGRAMMING USING FUZZY MODELS

Fuzzy mathematical programming is a concept which is capable of modeling and handling problems described either by a crisp relation or by a fuzzy relation. This concept can also be used for solving multi-objective models with reasonable attempts [35]. Some essential classes of fuzzy mathematical programming are as follows.

1.5.1   Fuzzy Linear Programming

Fuzzy linear programming (FLP) methods were first introduced for the solution of linear programming problems, with the slight difference that the parameters were fuzzy instead of crisp numbers. Fuzzy linear programming is ideal for solving linear programming problems with various objective functions [36]. The main benefit of using FLP is that they are numerically very time saving and can be implemented in various ways in case of different decision actions and perspectives. Before jumping onto the FLP, it is important to know what actually objectives and feasibility mean. Feasibility is defined as the region that satisfies all constraints [30]. The objective is the solution in the feasibility region which gives the highest utility. In a fuzzy decision problem, the objectives and constraints are also represented by a fuzzy set [37]. Feasibility region or region of solutions is also a fuzzy set. The set of acceptable solutions is given as the intersection of all fuzzy sets comprising all constraints and objectives:

$$\mu_f(x) = \min\big(\mu_i(x)\big) \tag{1.3}$$

To maximize the decision

$$\max\big(\mu_f(x)\big) = \max\min\big(\mu_i(x)\big) \tag{1.4}$$

$\mu_i(x)$ represents all the fuzzy sets which represent constraints and objectives.

1.5.2 FUZZY INTEGER LINEAR PROBLEMS

In linear programming, we perform optimizations using linear functions by satisfying linear equalities, inequalities and constraints. Unfortunately, in the actual situation, the knowledge and information that we consider are not precise and accurate, which further leads to the development of FLP. Fuzzy system of linear equations contains fuzzy right-hand side. To solve fuzzy integer linear problems firstly, they are converted into integer linear problem. In FLP, the variables are limited to integer values [38]. The authors discussed three models to convert the fuzzy integer linear problems into simple integer learning problems.

1.5.3 FUZZY DYNAMIC PROGRAMMING

Fuzzy dynamic programming deals with the multi-variate and multi-stage decision making. Fuzzy dynamic programming deals with the ambiguity that is present in optimization problems. Common optimizing control problems are usually solved using fuzzy dynamic programming problems. In order to solve them, the technique that is used is that they are first solved using traditional dynamic programming approach and then fuzzy dynamic programming is applied to them. The three main components with which fuzzy dynamic programming deals are fuzzy goals, fuzzy decisions and fuzzy constraints. It does not matter whether the model is deterministic or stochastic; fuzziness can be formulated in existing decision models [39].

1.5.4 MATHEMATICAL PROGRAMMING AS A TOOL FOR FUZZY RULE LEARNING PROCESS

The authors in [40] used mathematical programming as a tool for the fuzzy rule learning process. First, they used fuzzy clustering and then designed an objective function different from fuzzy c-means clustering and generated clusters. Using these

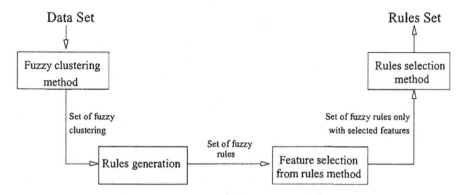

FIGURE 1.3 Fuzzy rule learning process [28].

clusters, they produced fuzzy rules where antecedent was a fuzzy cluster, and the terminus was the class with which cluster was labeled [28]. These rules helped in feature selection. Mathematical programming problem was designed to select features using this technique. They removed all unnecessary rules. A fuzzy mathematical problem was created to get knowledge about the rules that are vital for the classification task. In total, three mathematical programming problems were implemented for the fuzzy rule learning process (Figure 1.3).

1.6 MCDM WITH FUZZY AHP

To deal with complex decision problems, a powerful and resilient MCDM tool was created, which is known as the analytic hierarchy process (AHP) method. Under this method, a complicated system is divided into a hierarchical system of different elements. This hierarchical system usually contains the objectives, evaluation criteria and any other alternative for the solution to the problem. Various evaluation criteria may be present in the the evaluation criterion level. This level could also be further extended into a multi-layer architecture. All of the assessment parameters, mentioned in the associated level, are weighted to assess the final scores of the alternatives. After the final scores are determined, pairwise comparisons of different elements mentioned in each hierarchy are carried out using a nominal scale. After that, a comparison matrix is established by quantifying the correlations formed. The matrix's eigenvector is then computed. This vector measures the estimated weight of the various elements of a given hierarchy [41]. Finally, the eigen value is used to determine the strength of accuracy, comparative matrix ratio, and whether a piece of knowledge is suitable or not.

1.6.1 ANALYTICAL HIERARCHY PROCESS

AHP aims at capturing experts understanding of process under study. It comprises suggestions of experts along with the multi-criteria assessment [42]. It is a kind of MCDM technique popularly utilized in evaluating the decisiveness of the process. It formulates ratio scales by comparing the pairs of samples. AHP is a very simplified way and divides the job into a hierarchy of substructures [43]. Ranking of each

FIGURE 1.4 AHP hierarchy.

alternative is assigned in terms of relative significance for each criterion concerning other criteria and share to the approach being considered. Crisp values are then used to represent the comparison taken in the form of paired samples. A comparison matrix is formed by reiterating the process for each criterion. The obtained comparison matrix can be further used to calculate the weight vector. AHP relies on priority theory. It handles the complicated problems that involve the simultaneous cognizance of multi-criteria [41]. The hierarchical structure of the AHP is shown in Figure 1.4.

1.6.2 NEED FOR FUZZY AHP

Less effectiveness of simple AHP method to deal with the uncertainty in the decision-making process led to the growth of fuzzy AHP methods. The traditional AHP technique takes into account the definite opinion of decision-makers, thus making the process of comparison more adaptable and competent to explain expert's choices. At its most basic form, AHP evaluates the relative worth of decisive factors by contrasting decisions represented as crisp values. But, in most situations, human preference is dubious, and decision-makers seem to be more secure in harnessing linguistic variables rather than conveying their assessment in the form of numerical values. The fuzzy set theory [2] was integrated with AHP, to work with decision-making problems in a realistic scenario. Since fuzzy AHP is an extended version of AHP, it can deal with the stratified fuzzy decision-making jobs [44]. The evolution of fuzzy AHP method has led to its extensive usage by many researchers to deal with various problems of decision making in different areas, including excerption, assessment, asset allocation, development and planning. Fuzzy AHP model, being a prolongation of the AHP model, computes a series of weight vectors relying upon fuzzy sets to merge the pairwise scores of every feature, which is the average of these fuzzy scores [45]. The uncertainty linked with the decision-making task is easily handled by this model [21].

1.6.3   CASE STUDY OF FUZZY AHP

There exist various AHP methods that have incorporated the concept of fuzzy sets within them. The initial work in this domain was proposed by Ref. [42]. They implemented the logarithmic least-square approach to find triangular fuzzy pairwise comparison matrix and hence inferred fuzzy scores and weights. In Ref. [33], the authors dealt with the imprecision using the comparison ratio relied upon trapezoidal fuzzy numbers. They expanded Saaty's AHP [42] and incorporated the geometric mean method to infer fuzzy scores and weights [45]. In Ref. [46], the authors suggested a new extent analysis methodology relying upon fuzzy triangular numbers in order to perform a pairwise comparison. In many modern applications of fuzzy AHP changes, extent analysis is widely used as this technique is similar to that of the conventional AHP. Moreover, it requires low computational capacity in implementation. In Ref. [18], the authors pointed out the merits and demerits of various fuzzy AHP methods and compared different fuzzy AHP methods to differentiate between them [13].

1.6.4   APPLICATION OF FUZZY AHP: AIR POLLUTION CONTROL

In Ref. [21], the authors used fuzzy AHP Spatial Modeling of Air Pollution to evaluate the air pollution situation in Trehan. In the first step, the inverse distance weighting (IDW) was applied to obtain the estimate of the pollutants' density all over the study area. In the next level, five mentioned criteria were divided into sub-criteria according to the air quality index. Thus, two stages were considered to assign the weights: (i) the fuzzy AHP approach was used to weight the sub-criteria based on the pairwise rankings of the experts, and this procedure was applied to rank the main criteria. (ii) By obtaining the final weights, an overlay function was implemented to achieve the final air pollution susceptibility map [47].

1.7   COMPARISON BETWEEN AHP AND FUZZY AHP

The main aim of the analytic hierarchy process is to select the best criteria from multiple criteria. This method helps the experts to choose a better substitute and taking better decisions from all other alternatives by comparing each alternative pairwise. Predictions which are not well versed can be easily predicted using fuzzy AHP [48]. It helps in making quantitative predictions, which makes the quantitative forecasting easy. Vagueness arises during the opinions where the consistency occurs among the substitutes or say alternatives. In fuzzy AHP, or say in fuzzy pairwise comparisons, the criteria which have the least importance are weighted as zero. Traditional AHP method does not allow the situation where the alternative can have zero weight. Still, the numerical weight is close to zero if a criterion is less critical as compared to all of the others [49]. The significant benefit of using fuzzy AHP is that it can very easily ignore the requirements which have the least importance, whereas conventional AHP cannot. By merely just ignoring the significance of least essential criteria, fuzzy AHP can allow the decision-maker to concentrate only on those alternatives which are playing crucial roles in making a decision and are very important [50]. Classical and fuzzy methods are not competitors at the same conditions. One should use conventional AHP method when the information is definite and specific, but in case of vague, uncertain information, one should opt for fuzzy AHP method. Subjective assessments and linguistic assessments are done in questionnaire forms. Every linguistic variable is provided with its numerical measure in the default (predefined) scale.

1.8 CONCLUSION

Air contamination has recently emerged as a huge problem for all. Pollution impacts all, so estimating air pollution concentrations is critical, both from a social and an industrial standpoint. The models that are designed, predicted and applied make use of a massive database, which allows for easy forecasting of different scenarios. These models are adaptable to various parts of the world. MCDM techniques help to make the right choice when multiple choices are available for the decision-maker. FMCDM techniques can be used to analyze the quantitative and qualitative data that are present with respect to air pollution and then providing the appropriate solution for the problem.

REFERENCES

1. Singh, H., Gupta, M. M., Meitzler, T., Hou, Z. G., Garg, K. K., Solo, A. M., & Zadeh, L. A. (2013). Real-life applications of fuzzy logic. *Advances in Fuzzy Systems*.
2. Zadeh, L. A., Klir, G. J., & Yuan, B. (1996). *Fuzzy Sets, Fuzzy Logic, and Fuzzy Systems: Selected Papers* (Vol. 6). World Scientific.
3. Hwang, C. L., & Masud, A. S. M. (2012). Multiple *Objective Decision Making—Methods* and *Applications: A State*-of-the-*Art Survey* (Vol. 164). Springer Science & Business Media, Berlin, Heidelberg.
4. Roostaee, R., Izadikhah, M., & Hosseinzadeh Lotfi, F. (2012). An interactive procedure to solve multi-objective decision-making problem: an improvment to STEM method. *Journal of Applied Mathematics, 2012*, 1–18.
5. Aruldoss, M., Lakshmi, T. M., & Venkatesan, V. P. (2013). A survey on multi criteria decision making methods and its applications. *American Journal of Information Systems, 1*(1), 31–43.
6. Hwang, C. L., & Masud, A. S. M. (1979). Basic concepts and terminology. In Hwang, C.-L., & Masud, A.S.M. (Eds.), *Multiple Objective Decision Making—Methods and Applications* (pp. 12–20). Springer, Berlin, Heidelberg.
7. Jato-Espino, D., Castillo-Lopez, E., Rodriguez-Hernandez, J., & Canteras-Jordana, J. C. (2014). A review of application of multi-criteria decision making methods in construction. *Automation in Construction, 45*, pp. 151–162.
8. Begam, S., Vimala, J., Selvachandran, G., Ngan, T. T., & Sharma, R. (2020). Similarity measure of lattice ordered multi-fuzzy soft sets based on set theoretic approach and its application in decision making. *Mathematics, 8*, 1255.
9. Nikolić, D., Milošević, N., Mihajlović, I., Živković, Ž., Tasić, V., Kovačević, R., & Petrović, N. (2010). Multi-criteria analysis of air pollution with SO_2 and PM 10 in urban area around the copper smelter in Bor, Serbia. *Water, Air, and Soil Pollution, 206*(1–4), 369–383.

10. Yadav, J. Y., Kharat, V., & Deshpande, A. (2011, October). Fuzzy description of air quality: a case study. In *International Conference on Rough Sets and Knowledge Technology* (pp. 420–427). Springer, Berlin, Heidelberg.

11. Vo, T., Sharma, R., Kumar, R., Son, L. H., Pham, B. T., Tien, B. D., Priyadarshini, I., Sarkar, M., & Le, T. (2020). Crime rate detection using social media of different crime locations and twitter part-of-speech tagger with brown clustering, 4287–4299.

12. Domańska, D., & Wojtylak, M. (2012). Application of fuzzy time series models for forecasting pollution concentrations. *Expert Systems with Applications*, *39*(9), 7673–7679.

13. Wang, Q., Dai, H. N., & Wang, H. (2017). A smart MCDM framework to evaluate the impact of air pollution on city sustainability: a case study from China. *Sustainability*, *9*(6), 911.

14. Corani, G. (2005). Air quality prediction in Milan: feed-forward neural networks, pruned neural networks and lazy learning. *Ecological Modelling*, *185*(2–4), 513–529.

15. Nguyen, P. T., Ha, D. H., Avand, M., Jaafari, A., Nguyen, H. D., Al-Ansari, N., Van Phong, T., Sharma, R., Kumar, R., Le, H. V., Ho, L. S., Prakash, I., & Pham, B. T. (2020). Soft computing ensemble models based on logistic regression for groundwater potential mapping. *Applied Science*, *10*, 2469.

16. Tecer, L. H. (2007). Prediction of SO_2 and PM concentrations in a coastal mining area (Zonguldak, Turkey) using an artificial neural network. *Polish Journal of Environmental Studies*, *16*(4), 633–638.

17. Lad, R. K., Christian, R. A., & Deshpande, A. W. (2008). A fuzzy MCDM framework for the environmental pollution potential of industries focusing on air pollution. *WIT Transactions on Ecology and the Environment*, *116*, 617–626.

18. Büyüközkan, G., & Çifçi, G. (2012). A combined fuzzy AHP and fuzzy TOPSIS based strategic analysis of electronic service quality in healthcare industry. *Expert Systems with Applications*, *39*(3), 2341–2354.

19. Sowlat, M. H., Gharibi, H., Yunesian, M., Mahmoudi, M. T., & Lotfi, S. (2011). A novel, fuzzy-based air quality index (FAQI) for air quality assessment. *Atmospheric Environment*, *45*(12), 2050–2059.

20. Jha, S. et al. (2019). Deep learning approach for software maintainability metrics prediction. *IEEE Access*, *7*, 61840–61855.

21. Sheikhian, H. (2015). Spatial modeling of air pollution in urban areas applying fuzzy-AHP method; a case study of Tehran, Iran. Doi: 10.13140/RG.2.1.4146.1920.

22. Hájek, P., & Olej, V. (2009). Air pollution assessment using hierarchical fuzzy inference systems. Scientific Papers of The University Of Pardubice, 52.

23. Carbajal-Hernández, J. J., Sánchez-Fernández, L. P., Carrasco-Ochoa, J. A., & Martínez-Trinidad, J. F. (2012). Assessment and prediction of air quality using fuzzy logic and autoregressive models. *Atmospheric Environment*, *60*, 37–50.

24. Gurgatz, B. M., Carvalho-Oliveira, R., de Oliveira, D. C., Joucoski, E., Antoniaconi, G., do Nascimento Saldiva, P. H., & Reis, R. A. (2016). Atmospheric metal pollutants and environmental injustice: a methodological approach to environmental risk analysis using fuzzy logic and tree bark. *Ecological indicators*, *71*, 428–437.

25. Abbasi, F. (2017). The fuzzy logic in air pollution forecasting model. *International Journal of Industrial Mathematics*, *9*(1), 39–45.

26. Vahdat, S. E., & Nakhaee, F. M. (2011). Air pollution monitoring using fuzzy logic in industries. In *Advanced Air Pollution*. InTech.

27. Sharma, R., Kumar, R., Sharma, D. K., Son, L. H., Priyadarshini, I., Pham, B. T., Bui, D. T., & Rai, S. (2019). Inferring air pollution from air quality index by different geographical areas: case study in India. *Air Quality, Atmosphere, and Health*, *12*, 1347–1357.

28. Hamedian, A. A., Javid, A., Zarandi, S. M., Rashidi, Y., & Majlesi, M. (2016). Air quality analysis by using Fuzzy inference system and Fuzzy C-mean clustering in Tehran, Iran from 2009–2013. *Iranian Journal of Public Health*, *45*(7), 917–925.

29. Alhanafy, T. E., Zaghlool, F., & Moustafa, A. S. E. D. (2010). Neuro fuzzy modeling scheme for the prediction of air pollution. *Journal of American Science*, *6*(12), 605–616.

30. Sharma, R., Kumar, R., Singh, P. K., Raboaca, M. S., & Felseghi, R.-A. (2020). A systematic study on the analysis of the emission of CO, CO_2 and HC for four-wheelers and its impact on the sustainable ecosystem. *Sustainability*, *12*, 6707.

31. Vargas, L. G. (1990). An overview of the analytic hierarchy process and its applications. *European Journal of Operational Research*, *48*(1), 2–8.

32. Amini, R., & Ng, S. C. Comparison of Artificial Neural Network, Fuzzy Logic and Adaptive Neuro-Fuzzy Inference System on Air Pollution Prediction. No. table of contents pages 1 backdoor detection using machine learning, 14.

33. Juhos, I., Makra, L., & Tóth, B. (2008). Forecasting of traffic origin NO and NO_2 concentrations by Support Vector Machines and neural networks using Principal Component Analysis. *Simulation Modelling Practice and Theory*, *16*(9), 1488–1502.

34. Sharma, S. et al. (2020). Global forecasting confirmed and fatal cases of COVID-19 outbreak using autoregressive integrated moving average model. *Frontiers in Public Health*. https://doi.org/10.3389/fpubh.2020.580327.

35. Zimmermann, H. J. (1983). Fuzzy mathematical programming. *Computers & Operations Research*, *10*(4), 291–298.

36. Jarray, F. (2011). Discrete tomography and fuzzy integer programming. *Iranian Journal of Fuzzy Systems*, *8*(1), 41–48.

37. Allahviranloo, T., Shamsolkotabi, K. H., Kiani, N. A., & Alizadeh, L. (2007). Fuzzy integer linear programming problems. *International Journal of Contemporary Mathematical Sciences*, *2*(4), 167–181, Zahedan, Iran.

38. Herrera, F., & Verdegay, J. L. (1995). Three models of fuzzy integer linear programming. *European Journal of Operational Research*, *83*(3), 581–593.

39. Nieto-Morote, A., & Ruz-Vila, F. (2011). A fuzzy AHP multi-criteria decision-making approach applied to combined cooling, heating, and power production systems. *International Journal of Information Technology & Decision Making*, *10*(03), 497–517.

40. Cadenas, J. M., Garrido, M. C., & Hernandez, J. J. (2001). Fuzzy modelling using fuzzy mathematical programming. In *EUSFLAT Conference* (pp. 10–13).

41. Do, Q. H., & Chen, J. F. (2014). Group MCDM based on the fuzzy AHP approach. In Do, Q. H., & Chen, J. F. (Eds.), Encyclopedia of Business Analytics and Optimization (pp. 1100–1106). IGI Global.

42. Van Laarhoven, P. J. M., & Pedrycz, W. (1983). A fuzzy extension of Saaty's priority theory. *Fuzzy Sets and Systems*, *11*(1–3), 229–241.

43. Dansana, D. et al. (2021). Using susceptible-exposed-infectious-recovered model to forecast coronavirus outbreak. *Computers, Materials & Continua*, *67*(2), 1595–1612.

44. Vo, M. T., Vo, A. H., Nguyen, T., Sharma, R., & Le, T. (2021). Dealing with the class imbalance problem in the detection of fake job descriptions. *Computers, Materials & Continua*, *68*(1), 521–535.

45. Sachan, S., Sharma, R., & Sehgal, A. (2021). Energy efficient scheme for better connectivity in sustainable mobile wireless sensor networks. *Sustainable Computing: Informatics and Systems*, *30*, 100504.

46. Chang, D. Y. (1996). Applications of the extent analysis method on fuzzy AHP. *European Journal of Operational Research*, *95*(3), 649–655.

47. Kapageridis, I., Evagelopoulos, V., & Triantafyllou, A. (2009). Prediction of PM10 concentrations using a modular neural network system and integration with an online air quality management system.

48. Malik, P. et al. (2021). Industrial internet of things and its applications in industry 4.0: state-of the art. *Computer Communication*, *166*, 125–139, Elsevier.

49. Ghanem, S. et al. (2021). Lane detection under artificial colored light in tunnels and on highways: an IoT-based framework for smart city infrastructure. *Complex & Intelligent Systems*. https://doi.org/10.1007/s40747-021-00381-2.
50. (2020). Analysis of water pollution using different physico-chemical parameters: a study of Yamuna river. *Frontiers in Environmental Science*. https://doi.org/10.3389/fenvs.2020.581591.

2 Security and Privacy Requirements for IoMT-Based Smart Healthcare System
Challenges, Solutions, and Future Scope

Mandeep Singh, Namrata Sukhija,
and Anupam Sharma
HMR Institute of Technology and Management

Megha Gupta
IMS

Puneet Kumar Aggarwal
ABES

CONTENTS

DOI: 10.1201/9781003032328-2

2.1 INTRODUCTION

The numbers of patients are constantly increasing along with the time. To provide evasion of sicknesses, effective steps must be taken to improve the significance of healthcare sector. Mitigation is not just done by normal exercise, nourishment, and intermittent deterrent controls, as an approach to support a better environment yet additionally we must include some strategy for holding significant conditions back from getting poorer. Patients with the critical conditions should be given the prior consideration for their treatments; we should require a new approach to remove the pressure from the healthcare systems [1].

In the healthcare industry, critical upgrades in proficiency and guidelines/standards are implemented to improve the scope of improvements in Internet of Medical Things (IoMT). With the advancements in the field of microelectronics, there is an increased interest in the smart wearable medical gadgets in the IoHT or IoMT industry [2]. The fast advancement of IoHT, nonetheless, has implied protection and security in IoHT-based medical care frameworks frequently has gotten deficient consideration. Future medical or healthcare industry should have the scope of handling the increasing number of static cases and the shortfall in medicines to satisfy the requests [3]. Corona virus has featured the significance of speedy, extensive, and precise eHealth-care, besides, wise medical care including various sorts of clinical records to analyze the infection [4].

In smart medical framework, Internet of Things (IoT) enhanced the healthcare in terms of data accessibility, traceability, and maintainability of data by leverage of gadgets. Various wearable IoT-based gadgets are implanted in smart healthcare, and these gadgets utilize ad-hoc network to effortlessly access healthcare reports and connect individuals, assets, and organizations [5]. Smart medical treatment incorporates assorted actors, including doctors, staff, medical clinics, and exploration bodies. It includes on-demand system with numerous aspects, identification of diseases and prevention, the executives of medical care, and clinical survey [6,7].

In our lives through technology, automation, innovation, and robotized, various kinds of sensors are moderately implanted in our distributed nature system. Data generated by sensors empower the medical care framework to take the rapid action in critical circumstances reliably and assist the dependent for upcoming treatments in advance [8].

IoT gradually begins to interface between the specialists and customers through medical services. Electroencephalogram (EEGs), blood pressure (BP) readings, Electrocardiography (ECGs), glucose receptors, ultrasounds, and more keep on observing patients' health [9]. In case of critical conditions, patients need to visit the specialist. Smart beds are being used in the medical care framework for continually monitoring the patient's vitals and changing the position of bed according to the need. In medical care framework, IoT utilizes the IoMT to provide facilities to the patients. Fusion of IoMT and healthcare can play important role to providing the facilities for users [10,11].

The current scenario, as numerous kinds of infections and viruses are coming into existence, depending on the same kind of medical signals may not be sufficient to fulfill the prerequisites for analyzing them. Different medical signals and signs can be combined at various levels [12].

At the point when these signs consolidated various troubles may be capable and challenges fuse in synchronization when acquiring signals from different order mixes, include standardization, information buffering, and sensor [13,14].

The IoT has been generally recognized as a solution for reducing the pressing factors on the healthcare system and has been the focal point of recent research [15]. A lot of this exploration takes a gander at checking patients with specific conditions like diabetes [16] or Parkinson's disease [17]. Further research looks to fill explicit needs, for example, helping recovery through steady observing of a patient's improvement [18].

2.1.1 MOTIVATION FOR THE CHAPTER

This chapter consequently makes an exceptional commitment in that it distinguishes all critical parts of the IoT medical services framework and proposes a conventional model that could be applied to all IoT-based medical services frameworks (see Figure 2.1). This is essential as there is still no known system for distant checking of health in the literature. It provides a comprehensive survey of the state-of-the-art IoT-based smart healthcare framework. This research is majorly focusing on sensors for checking different health criteria, short and long-range correspondence rules, likewise on cloud innovations. It separates itself from the past significant overview commitments by considering each fundamental component of an IoT-based healthcare system as a framework as well as independently [19].

This chapter utilizes bottom-up methodology, inspecting the privacy challenges, security and important aspects about the patient's information to the healthcare level in IoMT-based medical care frameworks. Further, it also provides the biometric devices and sensors the capability to IoMT medical services frameworks. Additionally, it talks about the security plans for wearable IoMT gadgets, as the numbers of wearable gadgets are expending in smart medical care and listed the remarkable difficulties because of their equipment constraints.

2.1.2 ORGANIZATION OF THE CHAPTER

The rest of the chapter is organized into different sections. Section 2.2 discusses about the concept of the medical scenarios, IoMT, cloud- and edge-based intelligent

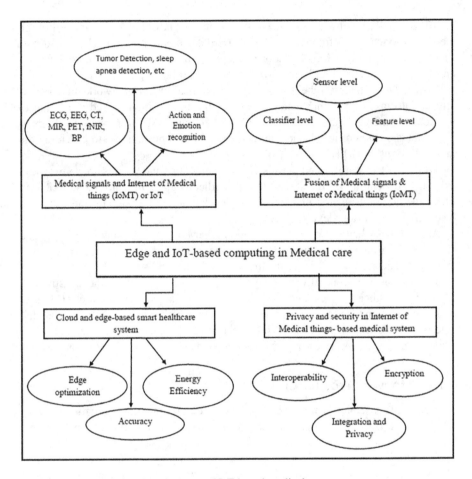

FIGURE 2.1   Classification of edge- and IoT-based medical care.

and smart medical systems and how it is useful for the healthcare system. Section 2.3 discusses about the privacy and security requirements for IoMT medical care frameworks and how they are important for individual as well as for the healthcare system. Section 2.4 illustrates various security approaches for IoMT healthcare systems along with future research directions and how it will help to improve the security aspects of the system, and in Section 2.5, conclusions are drawn from the chapter. Finally, references are listed as the last section.

2.2   RESEARCH AREAS

This survey is categorized in to four parts: medical signals and IoMT or IoT; fusion of medical signals and IoMT; edge- and cloud-based smart and intelligent medical or healthcare system (see Figure 2.1); and privacy and security in IoMT-based healthcare system.

2.2.1 Internet of Medical Things or Internet of Things and Medical Signals

Miao et al. [20] utilized a multi-sensor approach of Pressure Pulse Wave (PPW) with two channel signs and to assess BP with one-channel ECG. The signs are gathered from a sum of 35 physiological as well as instructive attributes that were extricated. The measurement decreases and the most encouraging criterion for each subject to acquire the selection strategy based on weakly supervised feature (WSF) introduced by them. Besides, a regression algorithm with multi-examples was utilized to intertwine attributes to improve the BP model.

In mine regions, the main threat is on the lives of the laborer. An ongoing observing framework proposed by Gu et al. [21] to make sure precision is in place diminishes the dangers to the mine labors. It also studies about multi-sensor data merging, circumstance cognizance, which includes the IoT. To consolidate the information and to determine the level of the circumstance, an SVM-based model as random forest (RF) was utilized. Root mean square error (RMSE) underneath 0.2 and a TSQ no more prominent than 1.691 after 200 emphases were shown by the recreation examination.

Steenkiste et al. [22] provide an authentic method to improve the presentation and authenticity of prediction of sleep apnea dependent on sensor strategy of fusion. Multi-sensor data can be assembled and facilitated, including oxygen submersion, heartbeat, stomach, and thoracic respiratory belt, the proposed approach used backward backup course of action affiliations [23]. To survey power and dissect the performance of the proposed combination strategy, both long- and short-term memory (LSTM) as well as deep learning base models as convolutional neural network (CNN) were utilized.

Utilizing particle filtering, a methodology was proposed by Nathan and Jafari [24] to modify the tracking system of pulse existing remnant and the utilization of sensors which are wearable. They assessed the pulse separated from other sign attributes and to utilize the consistent, also designed the procedure of observation. Combination of information has been further from totally different signal modalities and sensors to expand the accuracy of varied observation. The proposed approach of improved performance was analyzed on moving objects brought about by Points Per Game (PPG) and ECG information in comparison to the accelerometer perceptions, and the outcome showed empowering error levels normally up to two beats per minute [25].

2.2.2 Fusion of Medical Signals and Internet of Medical Things

Swayamsiddha and Mohanty [26] examined the various different cognitive IoMT (CIoMT) to handle the pandemic of COVID-19 for various utilizations. The audit showed that the study of CIoMT was an effective and successful approach for easy recognition, diminishing the responsibility of the healthcare services, timely follow-up, and dynamic checking [27].

Yang et al. [28] explained a mixture of medicine which takes point-of-care for IoMT to examine patients in obtaining applicable clinical thought reception. The proposed stage may screen illness spread and furthermore decline the general medical services cost at public level. Singh et al. [29] included the general utilizations of the IoT reasoning in dealing with the COVID-19 pandemic crisis. This investigation is intended to scale back prices and modify the results of treatment by using different linked associations for the exchange of knowledge and productive stream [30].

In Zhang et al. [31], a fusion tree-routing algorithm was introduced which is an event-driven information. This article of ours discusses about the data gathering framework and the theory of healthcare data, which is separated into two main structures as client management frameworks and terminal hubs. Visual methods of IoT communication are the proposed approach of design for communication and modeling according to the various medical attributes. The outcome showed a redesign in exactness and idealness contrasted in various methods. Dimitrov [32] presented a novel methodology of Big Data and IoMT applications in the clinical benefits that permit inventive plans of action and considered assortments in work movement, patient's input, and yield upgrades. Adaptable applications and wearable sensors were in practice to comply different medical services which needs to assemble Big Data from patients to push medical services tutoring.

Luna-delRisco et al. [33] kept an eye on affirmation, obstacles to execution, and have a dread of the usage of advancement that is wearable in the Latin American clinical medical services system. The creators noticed some difficult issues that incorporated the plan and designation of HR for clinical offices, the accessibility of public thought, sponsoring fundamental courses of action for medical care projects, which offers divergence in medical services. The important arrangement considered was the smart wearable sensors [34].

Hossain et al. [35] introduced for seizure discovery a deep CNN model using a cross-patient seizure classifier phenomenal. Representation strategy exhibits that the characteristics learned by the CNN in different recurrence groups are the spatial distribution when examining the various classes.

Many survey papers on medical signals and IoMT (e.g. Joyia et al. [36]) introduced the IoT commitments in medical or healthcare fields with many significant issues regarding IoMT. Various research and applications in IoMT were talked about regarding how they tackled challenges faced by the worldwide medical service branches. Irfan and Ahmad [37] re-evaluate the present engineering model scenario to design a different and new model for IoHT or IoMT. These medical models specifically marked the inspirations that would inspire clinical professionals for choosing to espouse the IoMT and hence represented security and privacy issues related to IoMT. Al-Turjman et al. [38] introduced a far-reaching audit of the many recent engineering gadgets for IoMT and studied the various parts of IoMT, including various modules of communication significant technologies of sensing. In this section discussed about all the challenges and opportunities related to usage of IoMT for the medical services industry. The maim components of the IoMT infrastructure are data acquisition, cloud server, and the communication gateways [39].

2.2.3   Cloud- and Edge-Based Smart Healthcare System

Hossain and Muhammad [40] proposed a framework for the emotion recognition using a CNN, and for security and privacy insurances, it used the cloud- and edge-based programming. In Ref. [41], energy efficiency smart-healthcare framework captivated with various methods based on fuzzy classification which leads to interior seizure recognition. The Mobile Health Cloud (MHC) was being communicated with raw EEG information that was ready at the edge. The planned energy-restricted framework gets exhausted by decreasing the quantity of communicated information which gave high classification exactitude. The output provides expansion of battery lifetime up to 60% and a classification exactitude on top of 98. A current organization model, CIoT, has been arranged dependent on the utilization of scholarly registering advancements [42].

Zhang et al. [43] combined with the numerous choices of cloud, edge registering and mental component extraction to characterize a shrewd clinical consideration administration system through the edge-psychological processing based (ECC-based) which will allot a shiny new edge to the upper danger patients exploitation the best edge figuring assets. Various observational examinations focused on the proposed framework which was equipped with improved energy productivity with quality of experience of many clients. The authors in Ref. [44] introduced a computing design as edge-IoMT that reduced idleness and improvised transfer speed efficiency to a minimum level. It comprised mainly two technologies: cloud framework modules which safely communicate clinical data to the doctor, and the modules of edge computing which provide the compressed and filtered real-time video information [45].

In Ref. [46], Kharel et al. worked and used the remote communication termed as utilized Long Range (LoRa) and Fog computing to introduce a framework for smart and remote healthcare monitoring. Communication over long ranges with various energy utilization methods for different IoT gadgets and utilized within a studied framework to connect the cloud or edge client's gadget in a medical or healthcare center is done with LoRa radio. FC maintains transfer speed of networks and diminishes idleness for limiting information trade through the cloud. Many tests demonstrated that LoRa and Fog computing had very encouraging performance in medical services monitoring in remote areas [47].

In Ref. [48], the designer used various sensors which are wearable with a Deep Learning (DL) strategy (in particular a neural network [RNN] that is recurrent in nature) to present action forecast framework for humans. Features, data, and action expectation were prepared on quick edge gadgets like PCs. To foresee human exercises from an easily opened data set, i.e., public, leads to the Recurrent Neural Network (RNN) which is dependent on the characteristics, accomplishing a mean of 99% expectation reliability, and hence, the performance increases [49].

The sufficiency of Big Data by investigating or proposing partner model of edge-based treatment using hub to see the profound neural organization (DNN), Chung and Yoo [50]. Server model with an edge-based healthcare model was set up independently to handle all the response latency issues. Now, the outcomes represented that fusion of parallel processing models and DNN techniques minimized response time delay. Table 2.1 presents the summary of papers on the medical signals and IoMT or IoT, fusion of medical signals and IoMT, and edge- and cloud-based intelligent and smart industry of medicare or healthcare [51].

2.2.4 SECURITY AND PRIVACY IN INTERNET OF MEDICAL THINGS-MEDICAL OR HEALTHCARE-BASED SYSTEM

In smart medical services, a patient's clinical information should be maintained with secrecy because these records carry sensitive data about the patient. On the off chance that the protection and security compromised from medical services structure, that may be annoyed the patient in the shared. If a sports person's personal medical record is disclosed to public, then the opponent team may utilize those record to get illicit benefits. So, clinical information ought to be managed secretly, and the information should be transferred securely over the network [52].

Alsubaei et al. [53] introduced a scientific categorization of privacy and safety in the IoMT. They classified IoT layers (middleware, application, network, business, perception), interloper types (individual, organized team, state-supported team), impact (compromise of brand esteem, information divulgence, life risk), and attack strategy (malware, hardware and software bugs, implementation layer, social engineering). The perception layer incorporates wearable gadgets, for example, respiratory sensors, BP sensors, and fitness trackers; implantable gadgets, for example, case cameras; ambient gadgets like daylight sensors and entryway sensors; and fixed gadgets like X-rays and CT scanners. While there are numerous approaches to combine data from these gadgets, those are not explained in this survey [54,55].

Alsubaei et al. [56] identified the security and privacy threats to the IoMT-based healthcare framework and provided the progression safety approaches to handle these threats. In this paper, various security issues are referenced; disregarding the built-in parts for security, customer is not aware of safety approaches, but instead they focus on advertising and business advantages; and there is no agreement exists between the partners for security parameters. The authors in Ref. [57] introduced the blockchain of IoMT for implementing security and privacy in advanced medical services system. They also proposed a four-layer Blockchain for Internet of Medical Things architecture [58].

Alsubaei et al. [59] laid out an assessment framework that utilized the online platform to implement the security parameters in IoMT and listed the steps to follow for secure interaction between customer and patients. These web-based approaches also help to assess the healthcare and provide solution for the obstacles present in IoMT.

In Hamidi [60], smart medical services implement the biometric feature to enhance the security in the IoMT and provide the secure interface for transmission and access of data. In the current scenario, ID cards and passwords are used in the entry control section for security, but it can be immediately compromised and do not provide sufficient security to the sensitive medical data records. In this paper, the authors proposed a biometric based on the degree of safety procedures: control, communication, scientific, and IoMT gadget [61].

TABLE 2.1
Summary of Papers on the Medical Signals and Internet of Medical Things (IoMT) or IoT, Fusion of Medical Signals and Internet of Medical Things (IoMT), and Edge- and Cloud-Based Smart Healthcare System

Reference	Year	IoMTs	Purpose	Description
Miao et al. [20]	2020	One ECG, pulse pressure wave sensor (PPWS)	Measurement of BP	Use a multi-sensor approach of PPW with two-channel signs and to assess BP with one-channel ECG
Gu et al. [21]	2017	Multi-sensor	Health and risk assessment for miners	Study about multi-sensor data merging, circumstance cognizance
Steenkiste et al. [22]	2019	Heart rate, oxygen saturation, abdominal respiratory belt	Sleep apnea detection	Deep learning base-models such as CNN and LSTM were utilized
Nathan and Jafari [24]	2018	Wearable sensor ECG, PPG	Heart rate estimation	Performance was analyzed on moving objects brought about by PPG and ECG information in comparison to the accelerometer perceptions
Swayamsiddha and Mohanty [26]	2020	Multi-sensor	Pandemic of COVID-19 handled by CIoMT	Approach for easy recognition, diminishing the responsibility of the healthcare services, timely follow-up, and dynamic checking
Dimitrov [32]	2016	Wearable sensors, Big Data	Various innovative commercial models were permitted in healthcare supported by Big Data	Wearable sensors and adaptable applications were used to fulfill different medical services
Hossain et al. [35]	2019	Health monitoring done with the help of human action and activity reorganization	Seizure detection and utilizing done with deep CNN model	CNN in various frequency bands for spatial distribution of the characteristics
Abdellatif et al. [41]	2019	The EEG wearable device	To provide swift in-network classification (SIC)	Emotion recognition done by using CNN, and for security and privacy insurances used the cloud- and edge-based programming
Kharel et al. [46]	2019	Energy-efficient wearable gadgets	Remote communication	Long-range communication and energy utilization for various IoT gadgets
Chung and Yoo [50]	2020	Smart sensor	Handle the response latency issues	Proposing associate different edge-based treatment model utilizing DNNs to see nodes

2.3 SMART HEALTHCARE SYSTEMS: PRIVACY AND SECURITY REQUIREMENTS

Requirement of privacy and security of any data is the most important concern for any researcher. In the healthcare systems, the security and privacy should be at a supreme level. For an IoMT healthcare infrastructure, privacy and reliability requirements are more tedious than the traditional structure. Some new and updated security features are incorporated in the IoMT healthcare system as device positioning, which is an additional contribution to the privacy of the system [62]. The IoMT healthcare system works at different levels, and each level is defined with different traits of functionalities having separate privacy and security needs. Here we will study and discuss all the levels separately to better understand the healthcare systems. With reference to General Data Protection Regulation (GDPR) and Health Insurance Portability and Accountability Act (HIPAA), the security of the data will be the point of discussion at all the levels of the IoMT healthcare system [63].

2.3.1 DATA LEVEL

2.3.1.1 Confidentiality

Patient health data must be collected and stored in accordance with legitimate and ethical security guidelines like GDPR and HIPAA, leading limited access to only specific specialized persons. To such details, appropriate precautions have to be in place to prevent data breaches, which must be implemented in order to protect the healthcare data confidentiality. The impact of such interventions cannot be overstated. Cyber criminals could sell your information on the black market, causing you to lose money. Patients would be subjected to not only a breach of privacy, but also a breach of confidentiality. There is also the possibility of financial and reputational damage. Article 5(e) of GDPR states that the patient's personal data should be safe and protected. Then again, how long tolerant information can be held, HIPAA, puts no restrictions. There is a vast difference between the medical service providers who comply with HIPAA and GDPR. HIPPA compliers can reveal patients' saved and protected health data or information (PHI) to different health providers without their permission, whereas GDPR compliers should get permission explicitly from the patients [64].

2.3.1.2 Integrity

Motivation behind the IoMT medical services frameworks is the integrity of the data; that is, during the transmission of data, it is not compromised in any way until it reaches the destination from the source [65]. A striker can access and change a patient's information by breeching normal transmission from a remote organization, which could prompt extreme ramifications in perilous cases. To ensure that the information has not been undermined, the ability expected and unapproved twists or controls of the information is the basic. Accordingly, suitable instruments of information should honesty be executed to forestall modification of moved information by pernicious assaults. Likewise, the security of the data set aside in the clinical specialists besides ought to be ensured, which implies the information cannot be tempered along. GDPR Article 5(d) states that clinical benefit suppliers need to make fundamental measures to keep patients' information exact and cutting-edge. It additionally requires incorrect individual information to be eradicated or rectified at the earliest opportunity. The GDPR moreover weights on the 'precision' of the data, allowing data owners request expert communities for the amendment of any off-base data, and the expert centers should respond to the sales inside a timetable month. Additionally, HIPAA requires clinical benefit suppliers to embrace essential measures to guarantee any PHI put away in the frameworks that cannot be modified without approval [66].

2.3.1.3 Availability

Administrations and information should be open when they are required to the important clients. Such administrations and information, given by the clinical workers and gadgets, will be out of reach if DoS assaults happen. Any difficulty in reaching information or administrations could prompt hazardous episodes, like unfit to give brief alert on account of a cardiovascular failure. Accordingly, to oblige the chance of accessibility misfortune, the medical application services should guarantee information accessibility to the clients and administrations. As indicated by Article 32 of the GDPR, clinical benefit suppliers can re-establish the accessibility and admittance to individual information in an ideal way, for example, embracing preventive safety efforts and countermeasures to the tasks of assaults [67]. Additionally, patients in the EU hold, according to Article 17 of the GDPR, the favored interest of the data held for the clinical advantage providers must be erased, which is known as 'Failed to remember alternative', in any case, HIPPA doesn't required such right [68].

2.3.2 SENSOR LEVEL

The privacy, security, and protection face more difficulties in the sensor level of IoMT medical services framework because of the restricted computational capacity and force requirement of the clinical gadgets and sensors [69]. Latest research in sensor-level security is to place a large portion of calculations in the individual worker level by considering all things, and the safety efforts in the levels of sensors need to be less correspondent and light-weight overheads.

2.3.2.1 Tamper-Proof Hardware

The gadgets of IoMT, particularly sensors of surrounding, that prompt secure and private data are to be uncovered for aggressors. Other than this, the taken devices might be remade and redeployed to the system by attackers and tuning to trades furtively [70]. Subsequently, actual burglary of clinical gadgets is a serious security danger; furthermore, it should be tended to in the IoMT medical services frameworks. Clinical gadgets in the frameworks ought to in any event have altered safe incorporated circuits and forestalling codes stacked on the gadgets being perused by outsiders whenever being sent. To use actually un-clonable capacities (PUFs) as a model plan for getting data to place in the incorporated circuits (ICs) of the clinical devices [71].

2.3.2.2 Localization

Two sorts of sensor repression, one the situation of the sensor on-body and sensor's/patient's region in the indoor environment, is the principle center for all researchers.

The previous sensor restriction is intended to distinguish between the sensors that are found in the ideal body positions. Identification of such sensor positions on body is of required significance for applications, for example, action acknowledgment [72]. Location of Things (LoT) additionally known as sensor limitation, which is planned for the area of the sensor in a room or to find the actual position of the patient who is wearing the sensor in a structure. Also, because of the plan of the IoMT medical care frameworks, clinical devices could move all through the association incorporation routinely. Accordingly, a real-time interruption recognition measure in an organization is required to permit sensors to leave and rejoin sporadically. A model of measures of such type is SVELTE [73], an interruption identification strategy that is going to report vindictive hubs which are going to join the organization to managers.

2.3.2.3   Self-Healing

Autonomic computing [74] includes self-mending, which is of incredible significance for the framework of IoMT, and the gadgets of IoMT will continue their activity after assaults on an organization. To accomplish self-recuperating, the framework of IoMT has the option to specify and analyze the assaults, and apply comparing security systems [75] with insignificant human intercession. Self-recovering systems passed on should moreover be light-weight, similarly as correspondence overheads to the association and computational multifaceted design to the clinical and clinical consideration contraptions. A self-mending design of model is proposed for IoT [76], which defines the specific calculation of all cells that is applicable within the organization that helps identifying assaults on a network. In various kinds of organization, the assaults require diverse recognition and strategies. Choosing different methods of autonomic security plans is very significant for network directors, which ought to be carried out for the organization.

2.3.2.4   Over the Air Programming

The refreshing (OTA) or absurd programming has become a famous strategy to refresh the framework of an IoT including an enormous count of various sensor hubs that present various privacy and security concerns; for example, all malevolent sensor centers refreshes creating characters in the association [77]. OTA can be important for self-recuperating instrument, refreshing security rules for the organization quickly. To carry out OTA appropriately, efforts for the safety should be designed to forestall OTA refreshes which are being abused in IoMT concept by aggressors. The arrangement of model is SEDA [78] that provides safe OTA programming convention dispersed for intended organization like framework of IoMT.

2.3.3   PERSONALIZED SERVER LEVEL

As we are aware that the patients' information is frequently put away and totaled in the individual worker level prior to being sent to the clinical workers in the IoMT medical care frameworks [79], it is fundamental to guarantee that the information is all around ensured while on the individual workers. For the most part, two kinds of confirmation plans should be sent to guarantee security and protection in the individual worker level, in particular gadget validation and client verification.

2.3.3.1 Device Authentication

Individual worker (e.g., a Personal Digital Assistant) will perform confirmation prior to tolerating data sent from the clinical contraptions additionally, sensors. Gadget confirmation plan ought to be capable to build up got/scrambled correspondences for information confidentiality and honesty [80].

2.3.3.2 User Authentication

Data sets are kept aside either by chance or everlastingly upon the individual specialists that ought to just be gotten to by the patients and clinical staff, like guards, in this manner, productive client certification plans are required [81].Particular workers in the IoMT clinical thought frameworks ought to in like way keeping up with the moment access of the information in the patients basic and fundamental conditions, for example, a seizure or a stroke.

2.3.4 MEDICAL SERVER LEVEL

Standard two necessities for the security in addition, protection of information identified with patients' data in the degree of clinical laborer, can be said as: just supported devices and workforce approach the data; and the genuine data ought to be encoded at record-breaking right when taken care of in the data bases [82].

2.3.4.1 Access Control

To guarantee just approved gadgets and staff approach to the clinical workers, control plans must be conveyed for powerful access. Request authorization or assent from a patient is very difficult each time to demand for the information access

As such, the expert associations of the clinical laborers should give specific access control for clinical patients, for instance, which data needs to pick which can be shared no assent is required and for which outcasts can approach. The mainstream arrangement of particular control of access is encryption based on quality, also termed as quality-based encryption (QBE) [83],which is arranged cryptography as open key where the keys from credits are strange keys (e.g., gotten signal strength, area, what's more, channel recurrence) [84].

2.3.4.2 Management of Trust or Faith

Trust suggests a two-path connection of two solid hubs, similar to a sensor center and organization coordinator, which share information or data with paths of one another. Likewise, creator in Ref. [85] examines about faith or trust to the degree up to which a hub is reached and definable when it helps out the other hub. Spread participation of the patient's information or data between the hubs of organization should be effective for remote medical care applications [86].

2.4 CHALLENGES AND FUTURE RESEARCH DIRECTIONS

Main and significant difficulties of IoT- and AI-based smart medical vigilance incorporate interoperability of sensors, privacy and security, gadget's data transmission, gadget management, data management obstacle, and efficient utilization of AI. In some medical conditions, IoMT gadgets in bulk can be utilized to recognize and analyze a disease; the gathered data from various heterogeneous sensors contain a large number of issues, like drained batteries, availability issues, or equipment glitches [87]. In medical healthcare, there are many basic issues that are unregulated till now. In particular, some unexplained bugs during the usage of clinical sensors which are well known in the clinical documented, as smartwatches and phones. Some additional complexities are regular in nature, for example, battery power, differentiations of specific physical characteristics, and discrepancy between the environments [88].

The clinical benefits system will get unsure data from the assorted sensors due to the forgetfulness from the specialists. Divided information can be taken or shammed by others. Various frequency ranges of IoTs could influence checking zones, and clients could provide false readings. Mark crashes and tag detuning ought to be amended, within sight metal/liquid effects and name arrangements [89]. The structure will get monotonous knowledge that should be re-defined. The wearable sensors are outfitted with batteries, bluetooth, and a number of connecting materials which were planned or decided to be in touch or contacted to the human skin. For security of patients or mankind, it is very basic to contemplate hurtfulness, ignitable materialistic things, and completely different components during the designing for medical usable wearable sensors. These medical-based wearable sensors also constrain body movement for example, belt worn at lower leg or at the midriff is agonizing particularly for youths and the seasoned patients. This is a test for designers to create wearable devices and those can consistently screen each of the imperative signs without bargaining the client's solace [90].

Self-tuning, self-detecting, and self-changes are the vital prerequisites of the programmed medical care framework [91,92]. As establishment, for example, sensor commotion and recording climate, IoT and blend of sensors can deal with the changes, seeing as they can straight-forwardly influence structure properties like exactness [93]. Data transmission procedures or strategies for move learning are to be utilized to permit the healthcare system to change as per explicit conditions by social affair and moving colleague from one situation then onto the following. Unapproved admittance to IoT devices might augment clinical and personal information dangers to patients. Related PCs, just as the recuperation, amassing, and crossing of patient data to the clinical or medical care information cloud. RF sticking, cloning, mocking, and reviewing of data cloud are helpless against structure sort. Inside the cloud study, traffic is engaged such a ton that orders will be mixed foursquare to a PC by an individual inside the affiliation [94].

Denials of service (DoS) attack may affect medical associations and the privacy and security of medical data of the patients. In spite of the way that replication (use of various devices in the association) is essentially a fixed or standard wellbeing and security of DoS, which is habitually not to be achievable to copy a portion of the utilization of assets in a clinical consideration design which determines that a bit of the devices are basic systems implant Inferable with the intricacy and total about the new medical care contraption and hardware bugs, indicates the expedient conspicuous confirmation of possible wellbeing and security hazards remains an issue. This issue is escalating as the Internet interfaces an ever-increasing number of clients [95]. Standard security is additionally far and wide today, and unstable UI access raises the threat surface more [96].

The pragmatic usefulness of IoMT-actuated medical services frameworks is seldom referred in literature. Main and fundamental thought of concern is that the most important information that is possessed by an organization is not open for the general public. Practically speaking, effective arrangement and use of information will take into consideration more solid estimation and assessment of everyday physical work using minimal effort that can prompt simpler and better preventive consideration for chronic diseases [97].

2.5 CONCLUSION

The framework for smart healthcare is a well-explored area. Smart medical services define that there is an expansiveness of literature covering AI, edge, IoMT, clinical signs, IoT, and distributed computing all at different rates and using many varied strategies. According to the study, till now we are lacking in the efficient and effective result of privacy and security aspects after the analysis of various methods such as IoT, IoMT, AI, usage of medical terms and fusion, cloud computing, and edge computing in the healthcare domain system.

Throughout the chapter, it has been discussed how, with the help of the IoMT, the healthcare system can be improved to provide better facilities to patients and also use medical data to analyze the various diseases and provide better results by continuously monitoring the patients with the help of medical gadgets.

The chapter focused on the subsequent comparison of all the strategies like edge, AI, IoMT, distributed computing, and IoT protection, privacy, and security in medical services based on the various classification methods. The study incorporated the utilization of IoT, clinical signs, and IoMT, as well as the combination of the utilization of edge and distributed computing and sensors in smart medical services. The chapter also focused on a wide range of safety and protection issues including all gadgets of IoMT. At long last, some research-based difficulties and future exploration bearings were talked about.

REFERENCES

1. Alshehri, F., & Muhammad, G. (2020). A comprehensive survey of the Internet of Things (IoT) and edge computing in healthcare. *IEEE Access* 9, 3660–3678
2. Sethi, R., Bhushan, B., Sharma, N., Kumar, R., & Kaushik, I. (2020). Applicability of industrial IoT in diversified sectors: Evolution, applications and challenges. *Studies in Big Data Multimedia Technologies in the Internet of Things Environment*, 45–67. doi: 10.1007/978–981-15-7965-3_4.
3. Reports & data. (2021, April 30). Australian Institute of Health and Welfare. http://www.aihw.gov.au/reports-data?id=60129548150.
4. Saxena, S., Bhushan, B., & Ahad, M. A. (2021). Blockchain based solutions to Secure Iot: Background, integration trends and a way forward. *Journal of Network and Computer Applications*, 103050. doi: 10.1016/j.jnca.2021.103050.
5. Banerjee, A., Chakraborty, C., Kumar, A., & Biswas, D. (2020). Emerging trends in IoT and big data analytics for biomedical and health care technologies. *Handbook of Data Science Approaches for Biomedical Engineering*, 121–152. doi: 10.1016/b978-0-12-818318-2.00005.

6. Alshehri, F., & Muhammad, G. (2021). A comprehensive survey of the Internet of Things (IoT) and AI-based smart healthcare. *IEEE Access*, *9*, 3660–3678. https://doi.org/10.1109/access.2020.3047960.

7. Gupta, A., Chakraborty, C., & Gupta, B. (2019). Monitoring of epileptical patients using cloud-enabled health-IoT system. *Traitement Du Signal*, *36*(5), 425–431. doi: 10.18280/ts.360507.

8. Goyal, S., Sharma, N., Bhushan, B., Shankar, A., & Sagayam, M. (2020). IoT enabled technology in secured healthcare: Applications, challenges and future directions. *Cognitive Internet of Medical Things for Smart Healthcare Studies in Systems, Decision and Control*, 25–48. doi: 10.1007/978-3-030–55833-8_2.

9. Amin, S. U., Alsulaiman, M., Muhammad, G., Mekhtiche, M. A., & Shamim Hossain, M. (2019). Deep learning for EEG motor imagery classification based on multi-layer CNNs feature fusion. *Future Generation Computer Systems*, *101*, 542–554. https://doi.org/10.1016/j.future.2019.06.027.

10. Gupta, A., Chakraborty, C., & Gupta, B. (2019). Medical information processing using smartphone under IoT framework. *Energy Conservation for IoT Devices*, 283–308. doi: 10.1007/978–981-13–7399-2_12.

11. Chakraborty, C., & Bhattacharya, S. (2020). Healthcare data monitoring under Internet of Things. *Green Computing and Predictive Analytics for Healthcare*, 1–18. doi: 10.1201/9780429317224–1.

12. Gope, P., & Hwang, T. (2016). BSN-care: A secure IoT based modern healthcare system using body sensor network. *IEEE Sensors Journal*, *16*(5), 1368–1376.

13. Zhu, N., Diethe, T., Camplani, M., Tao, L., Burrows, A., Twomey, N., Kaleshi, D., Mirmehdi, M., Flach, P., & Craddock, I. (2015). Bridging e-health and the Internet of Things: The SPHERE project. *IEEE Intelligent Systems*, *30*(4), 39–46.

14. Chakraborty, C., & Abougreen, A. (2018). Intelligent Internet of Things and advanced machine learning techniques for covid-19. EAI Endorsed Transactions on Pervasive Health and Technology, 168505. doi: 10.4108/eai.28-1-2021.168505.

15. Aggarwal, P. K., Jain, P., Mehta, J., Garg, R., Makar, K., & Chaudhary, P. (2021). Machine learning, data mining and big data analytics for 5G-enabled IoT. In Sudeep Tanwar (Ed.), *Blockchain for 5G Enabled IoT: The New Wave for Industrial Automation*, 351–375, Springer, New York.

16. Chang, S. H., Chiang, R. D., Wu, S. J., & Chang, W. T. (2016). A context-aware, interactive M-health system for diabetics. *IT Professional*, *18*(3), 14–22.

17. Pasluosta, C. F., Gassner, H., Winkler, J., Klucken, J., & Eskofier B. M. (2015). An emerging era in the management of Parkinson's disease: Wearable technologies and the Internet of Things. *IEEE Journal of Biomedical and Health Informatics*, *19*(6), 1873–1881.

18. Fan, Y. J., Yin, Y. H., Xu, L. D., Zeng, Y., & Wu, F. (2014). IoT based smart rehabilitation system. *IEEE Transactions on Industrial Informatics*, *10*(2), 1568–1577.

19. Banerjee, A., Chakraborty, C., & Rathi, M. (2020). Medical imaging, artificial intelligence, Internet of Things, wearable devices in terahertz healthcare technologies. *Terahertz Biomedical and Healthcare Technologies*, 145–165. https://doi.org/10.1016/b978-0-12-818556-8.00008-2.

20. Miao, F., Liu, Z.-D., Liu, J.-K., Wen, B., He, Q.-Y., & Li, Y. (2020). Multisensor fusion approach for cuff-less blood pressure measurement. *IEEE Journal of Biomedical and Health Informatics*, *24*(1), pp. 79–91.

21. Gu, Q., Jiang, S., Lian, M., & Lu, C. (2019). Health and safety situation awareness model and emergency management based on multi-sensor signal fusion. *IEEE Access*, *7*, 958–968.

22. Van Steenkiste, T., Deschrijver, D., & Dhaene, T. (2019). Sensor fusion using backward shortcut connections for sleep apnea detection in multi-modal data, *arXiv:1912.06879*. [Online]. Available: http://arxiv.org/abs/1912.06879.

23. Begam, S., Vimala, J., Selvachandran, G., Ngan, T. T., & Sharma, R. (2020). Similarity measure of lattice ordered multi-fuzzy soft sets based on set theoretic approach and its application in decision making. *Mathematics*, *8*, 1255.
24. Nathan, V., & Jafari, R. (2018). Particle filtering and sensor fusion for robust heart rate monitoring using wearable sensors. *IEEE Journal of Biomedical and Health Informatics*, *22*(6), 1834–1846.
25. Vo, T., Sharma, R., Kumar, R., Son, L. H., Pham, B. T., Tien, B. D., Priyadarshini, I., Sarkar, M., & Le, T. (2020). Crime rate detection using social media of different crime locations and twitter part-of-speech tagger with brown clustering, 4287–4299.
26. Swayamsiddha, S., & Mohanty, C. (2020). Application of cognitive Internet of medical things for COVID-19 pandemic. *Diabetes Metabolic Syndrome*, *14*(5), 911–915.
27. Nguyen, P. T., Ha, D. H., Avand, M., Jaafari, A., Nguyen, H. D., Al-Ansari, N., Van Phong, T., Sharma, R., Kumar, R., Le, H. V., Ho, L. S., Prakash, I., & Pham, B. T. (2020). Soft computing ensemble models based on logistic regression for groundwater potential mapping. *Applied Science*, *10*, 2469.
28. Yang, T., Gentile, M., Shen, C.-F., & Cheng, C.-M. (2020). Combining point of-care diagnostics and Internet of medical things (IoMT) to combat the COVID-19 pandemic. *Diagnostics*, *10*(4), 224.
29. Singh, R. P., Javaid, M., Haleem, A., and Suman, R. (2020). Internet of Things (IoT) applications to fight against COVID-19 pandemic. *Diabetes Metabolic Syndrome*, *14*(4), 521–524.
30. Jha, S. et al. (2019). Deep learning approach for software maintainability metrics prediction. *IEEE Access*, *7*, 61840–61855.
31. Zhang, Y., Zhang, Y., Zhao, X., Zhang, Z., & Chen, H. (2020). Design and data analysis of sports information acquisition system based on Internet of medical things. *IEEE Access*, *8*, 84792–84805.
32. Dimitrov, D. V. (2016). Medical Internet of Things and big data in healthcare. *Healthcare Informatics Research*, *22*(3), 156–163.
33. Luna-delRisco, M., Palacio, M. G., Orozco, C. A. A., Moncada, S. V., Palacio, L. G., Montealegre, J. J. Q., & Diaz-Forero, I. (2018). Adoption of internet of medical things (IoMT) as an opportunity for improving public health in Latin America, In *Proc. 13th Iberian Conf. Inf. Syst. Technol. (CISTI)*, Caceres, Spain, Jun. 2018, pp. 1–5.
34. Sharma, R., Kumar, R., Sharma, D. K., Son, L. H., Priyadarshini, I., Pham, B. T., Bui, D. T., & Rai, S., (2019). Inferring air pollution from air quality index by different geographical areas: Case study in India. *Air Quality, Atmosphere, and Health*, *12*, 1347–1357.
35. Hossain, M. S., Amin, S. U., Alsulaiman, M., & Muhammad, G. (2019). Applying deep learning for epilepsy seizure detection and brain mapping visualization. *ACM Transactions on Multimedia Computing, Communications, and Applications*, *15*(1s), 1–17.
36. Joyia, G. J., Liaqat, R. M., Farooq, A., & Rehman, S. (2017). Internet of medical things (IOMT): Applications, benefits and future challenges in healthcare domain. *Journal of Communication*, *12*(4), 240–247.
37. Irfan, M., & Ahmad, N. (2018). Internet of medical things: Architectural model, motivational factors and impediments, in *Proc. 15th Learn. Technol. Conf. (L&T)*, Feb. 2018, pp. 6–13.
38. Al-Turjman, F., Nawaz, M. H., & Ulusar, U. D. (2020). Intelligence in the Internet of medical things era: A systematic review of current and future trends. *Computer Communicatios*, *150*, 644–660.
39. Sharma, R., Kumar, R., Singh, P. K., Raboaca, M. S., & Felseghi, R.-A. (2020). A systematic study on the analysis of the emission of CO, CO_2 and HC for four-wheelers and its impact on the sustainable ecosystem. *Sustainability*, *12*, 6707.

40. Hossain, M. S., & Muhammad, G. (2019). Emotion recognition using secure edge and cloud computing. *Information Sciences*, *504*, 589–601.

41. Awad Abdellatif, A., Emam, A., Chiasserini, C.-F., Mohamed, A., Jaoua, A., & Ward, R. (2019). Edge-based compression and classiffication for smart healthcare systems: Concept, implementation and evaluation. *Expert Systems with Applications*, *117*, 1–14.

42. Banerjee, A., Chakraborty, C., Kumar, A., & Biswas, D. (2020). Emerging trends in IoT and big data analytics for biomedical and health care technologies. *Handbook of Data Science Approaches for Biomedical Engineering*, 121–152. https://doi.org/10.1016/b978-0-12-818318-2.00005-2.

43. Zhang, Y., Ma, X., Zhang, J., Hossain, M. S., Muhammad, G., & Amin, S. U. (2019). Edge intelligence in the cognitive Internet of Things: Improving sensitivity and interactivity. *IEEE Network*, *33*(3), 58–64.

44. Dilibal, C. (2020). Development of edge-IoMT computing architecture for smart healthcare monitoring platform, in *Proc. 4th Int. Symp. Multi- disciplinary Stud. Innov. Technol. (ISMSIT)*, Istanbul, Turkey, Oct. 2020, pp. 1–4.

45. Sharma, S. et al. (2020). Global forecasting confirmed and fatal cases of COVID-19 outbreak using autoregressive integrated moving average model. *Frontiers in Public Health*. https://doi.org/10.3389/fpubh.2020.580327.

46. Kharel, J., Reda, H. T., & Shin, S. Y. (2019). Fog computing-based smart health monitoring system deploying LoRa wireless communication. *IETE Technical Review*, *36*(1), 69–82.

47. Malik, P. et al. (2021). Industrial Internet of Things and its applications in industry 4.0: State-of the art. *Computer Communication*, *166*, 125–139, Elsevier.

48. Azeez, N. A., & Van der Vyver, C. (2018). Security and privacy issues in e-health cloud-based system: A comprehensive content analysis. *Egyptian Informatics Journal*, *20*(2), 97–108.

49. Sharma, S. et al. (2020). Analysis of water pollution using different physico-chemical parameters: A study of Yamuna river. *Frontiers in Environmental Science*. https://doi.org/10.3389/fenvs.2020.581591.

50. Chung, K., & Yoo, H. (2020). Edge computing health model using P2P based deep neural networks. *Peer-to-Peer Networking and Applications*, *13*(2), 694–703.

51. Dansana, D. et al. (2021). Using susceptible-exposed-infectious-recovered model to forecast coronavirus outbreak. *Computers, Materials & Continua*, *67*(2), 1595–1612.

52. Abou-Nassar, E. M., Iliyasu, A. M., El-Kafrawy, P. M., Song, O.-Y., Bashir, A. K., & El-Latif, A. A. A. (2020). DITrust chain: Towards block chain based trust models for sustainable healthcare IoT systems. *IEEE Access*, *8*, 111223–111238.

53. Alsubaei, F., Abuhussein, A., & Shiva, S. (2017). Security and privacy in the Internet of medical things: Taxonomy and risk assessment, in *Proc. IEEE 42nd Conf. Local Comput. Netw. Workshops (LCN Workshops)*, Singapore, Oct. 2017, pp. 112–120.

54. Bhushan, B., Sahoo, C., Sinha, P., & Khamparia, A. (2020). Unification of blockchain and Internet of Things (BIoT): Requirements, working model, challenges and future directions. *Wireless Networks*. doi: 10.1007/s11276-020-02445-6.

55. Jindal, M., Gupta, J., & Bhushan, B. (2019). Machine learning methods for IoT and their Future Applications, in *2019 Int. Conf. Comput. Commun. Intell. Syst. (ICCCIS)*, IEEE.

56. Alsubaei, F., Abuhussein, A., & Shiva, S. (2019). Ontology-based security recommendation for the internet of medical things. *IEEE Access*, *7*, 48948–48960.

57. Seliem, M., & Elgazzar, K. (2019). BIoMT: Blockchain for the Internet of medical things, in *Proc. IEEE Int. Black Sea Conf. Commun. Netw. (BlackSeaCom)*, Sochi, Russia.

58. Vo, M. T., Vo, A. H., Nguyen, T., Sharma, R., & Le, T. (2021). Dealing with the class imbalance problem in the detection of fake job descriptions. *Computers, Materials & Continua*, *68*(1), 521–535.

59. Alsubaei, F., Abuhussein, A., & Shiva, S. (2019). Ontology-based security recommendation for the internet of medical things. *IEEE Access, 7*, 48948–48960.
60. Hamidi, H. (2019). An approach to develop the smart health using Internet of Things and authentication based on biometric technology. *Future Generation Computer Systems, 91*, 434–449.
61. Sachan, S., Sharma, R., & Sehgal, A. (2021). Energy efficient scheme for better connectivity in sustainable mobile wireless sensor networks. *Sustainable Computing: Informatics and Systems, 30*, 100504.
62. Goyal, S., Sharma, N., Kaushik, I., & Bhushan, B. (2021). Blockchain as a solution for security attacks in named data networking of things. *Security and Privacy Issues in IoT Devices and Sensor Networks*, 211–243. doi: 10.1016/b978-0-12-821255-4.00010–9.
63. Ghanem, S. et al. (2021). Lane detection under artificial colored light in tunnels and on highways: An IoT-based framework for smart city infrastructure. *Complex & Intelligent Systems*. https://doi.org/10.1007/s40747-021-00381-2.
64. Mooney, G. (2018). Is HIPAA compliant with the GDPR? https://blog.ipswitch.com/is-hipaa-compliant-with-the-gdpr.
65. Pearlman, S. (2019). What is data integrity and why is it important? [Online]. Available: https://www.talend.com/resources/what-isdata-integrity/.
66. Ullah, K., Shah, M. A., & Zhang, S. (2016). Effective ways to use Internet of Things in the field of medical and smart health care, in *Proc. Int. Conf. Intell. Syst. Eng. (ICISE)*, Islamabad, Pakistan, Jan. 2016, pp. 372–379.
67. Bienkowski, T. (Feb. 2018). GDPR is explicit about protecting availability. [Online]. Available: https://www.netscout.com/blog/gdpravailability-protection.
68. Muzammal, M., Talat, R., Sodhro, A. H., & Pirbhulal, S. (2020). A multi-sensor data fusion enabled ensemble approach for medical data from body sensor networks. *Information Fusion, 53*, 155–164.
69. Sharma, N., Kaushik, I., Bhushan, B., Gautam, S., & Khamparia, A. (2020). Applicability of WSN and biometric models in the field of healthcare. In K. Martin Sagayam, Bharat Bhushan, A. Diana Andrushia, & Victor Hugo C. de Albuquerque (Eds.), *Deep Learning Strategies for Security Enhancement in Wireless Sensor Networks*, IGI Global Houston, TX.
70. Muhammad, L. J., Algehyne, E. A., Usman, S. S., Ahmad, A., Chakraborty, C., Mohammed, I. A. (2020). Supervised machine learning models for prediction of covid-19 infection using epidemiology dataset. *SN Computer Science, 2*(1). doi: 10.1007/s42979-020-00394-7.
71. Kephart, J. O., & Chess, D. M. (2003). The vision of autonomic computing. *Computer, 36*(1), 41–50.
72. Saeedi, R., Purath, J., Venkatasubramanian, K., & Ghasemzadeh, H. (2014). Toward seamless wearable sensing: Automatic on-body sensor localization for physical activity monitoring, in *Proc. 36th Annu. Int. Conf. IEEE Eng. Med. Biol. Soc.*, Aug. 2014, pp. 5385–5388.
73. Goyal, S., Sharma, N., Bhushan, B., Shankar, A., & Sagayam, M. (2020). IoT enabled technology in secured healthcare: Applications, challenges and future directions. *Cognitive Internet of Medical Things for Smart Healthcare – Studies in Systems, Decision and Control*, Springer International Publishing.
74. Singh, P. K., Rani, P., Samanta, D., Khanna, A., & Bhushan, B. (2020). An internet of health things-driven deep learning framework for detection and classification of skin cancer using transfer learning. Transactions on Emerging Telecommunications Technologies.
75. Stankovic, J. A. (2014). Research directions for the Internet of Things. *IEEE Internet Things Journal, 1*(1), 3–9.
76. de Almeida, F. M., de Ribamar Lima Ribeiro, A., & Moreno, E. D. (2015). An architecture for self-healing in Internet of Things, in *Proc. UBI-COMM*, p. 89.

77. Chen, M., Li, W., Hao, Y., Qian, Y., & Humar, I. (2018). Edge cognitive computing based smart healthcare system. *Future Generation Computer Systems, 86*, 403–411.

78. Kim, J. Y., Hu, W., Shafagh, H., & Jha, S. (2016). SEDA: Secure over the air code dissemination protocol for the Internet of Things. *IEEE Transactions on Dependable and Secure Computing, 15*(6), 1041–1054.

79. Bromwich, M., & Bromwich, R. (2016). Privacy risks when using mobile devices in health care. *Canadian Medical Association Journal, 188*(12), p. 855.

80. Crilly, P., & Muthukkumarasamy, V. (2010). Using smart phones and body sensors to deliver pervasive mobile personal healthcare, in *Proc. 6th Int. Conf. Intell. Sensors, Sensor Netw. Inf. Process.*, Dec. 2010, pp. 291–296.

81. Kogetsu, A., Ogishima, S., & Kato, K. (2018). Authentication of patients and participants in health information exchange and consent for medical research: A key step for privacy protection, respect for autonomy, and trustworthiness. *Frontiers in Genetics, 9*, 167.

82. Uddin, M. Z. (2019). A wearable sensor-based activity prediction system to facilitate edge computing in smart healthcare system. *Journal of Parallel and Distributed Computing, 123*, 46–53.

83. Han, T., Zhang, L., Pirbhulal, S., Wu, W., & de Albuquerque, V. H. C. (2019). A novel cluster head selection technique for edge-computing based IoMT systems. *Computer Networks, 158*, 114–122.

84. Jain, P., Aggarwal, P. K., Chaudhary, P., Makar, K., Mehta, J., & Garg, R. Convergence of IoT and CPS in Robotics. *Emergence of Cyber Physical Systems and IoT in Smart Automation and Robotics*, pp. 15–30, Springer.

85. Boukerche, A., & Ren, Y. (2009). A secure mobile healthcare system using trust-based multicast scheme. *IEEE Journal on Selected Areas in Communications, 27*(4), 387–399.

86. Rahman, M. A., Hossain, M. S., Islam, M. S., Alrajeh, N. A., & Muhammad, G. (2020). Secure and provenance enhanced Internet of Health things framework: A blockchain managed federated learning approach. *IEEE Access, 8*, 205071–205087. https://doi.org/10.1109/access.2020.3037474.

87. Al-Shargie, F. (2019). Fusion of fNIRS and EEG signals: Mental stress study. *engrXiv, 2019*, pp. 1–5. doi: 10.31224/osf.io/kaqew.

88. Khamparia, A., Singh, P. K., Rani, P., Samanta, D., Khanna, A., & Bhushan, B. (2020). An internet of health things-driven deep learning framework for detection and classification of skin cancer using transfer learning. *Transactions on Emerging Telecommunications Technologies*. doi: 10.1002/ett.3963.

89. Mishra, K. N., & Chakraborty, C. (2019). A novel approach toward enhancing the quality of life in smart cities using clouds and IoT-based technologies. *Internet of Things*, 19–35. doi: 10.1007/978-3-030-18732-3_2.

90. Bhushan, B., Khamparia, A., Sagayam, K. M., Sharma, S. K., Ahad, M. A., & Debnath, N. C. (2020). Blockchain for smart cities: A review of architectures, integration trends and future research directions. *Sustainable Cities and Society, 61*, 102360. doi: 10.1016/j.scs.2020.102360.

91. Bhushan, B., & Sahoo, G. (2020). Requirements, protocols, and security challenges in wireless sensor networks: An industrial perspective. *Handbook of Computer Networks and Cyber Security*, pp. 683–713. doi: 10.1007/978-3-030-22277-2_27.

92. Muhammad, G., Mesallam, T. A., Malki, K. H., Farahat, M., Alsulaiman, M., & Bukhari, M. (2011). Formant analysis in dysphonic patients and automatic arabic digit speech recognition. *BioMedical Engineering OnLine, 10*(1), 41.

93. Abou-Nassar, E. M., Iliyasu, A. M., El-Kafrawy, P. M., Song, O.-Y., Bashir, A. K., and El-Latif, A. A. A. (2020). DITrust chain: Towards blockchain based trust models for sustainable healthcare IoT systems. *IEEE Access, 8*, 111223–111238.

94. Kumar, S., Bhusan, B., Singh, D., & Choubey, D. K. (2020). Classification of diabetes using deep learning, in *2020 Int. Conf. Commun. Signal Process. (ICCSP)*, IEEE.
95. Yang, F., Zhao, X., Jiang, W., Gao, P., & Liu, G. (2020). Multi-method fusion of cross-subject emotion recognition based on high dimensional EEG features. *Frontiers in Computational Neuroscience*, *13*, 53. Accessed: Jul. 1, 2020. [Online]. Available: https://www.ncbi.nlm.nih.gov/pmc/articles/PMC6714862/.
96. Lin, K., Li, Y., Sun, J., Zhou, D., & Zhang, Q. (2020). Multi-sensor fusion for body sensor network in medical human–robot interaction scenario. *Information Fusion*, *57*, 15–26.
97. Harjani, M., Grover, M., Sharma, N., & Kaushik, I. (2019). Analysis of various machine learning algorithm for cardiac pulse prediction, in *2019 Int. Conf. Comput. Commun. Intell. Syst. (ICCCIS)*, pp. 244–249. doi: 10.1109/ICCCIS48478.2019.8974519.

3 The Rise of "Big Data" on Cloud Computing

Saurabh Singhal
GLA University Mathura, India

CONTENTS

3.1 INTRODUCTION

The beginning of social media can be accounted since 2004 when MySpace was the first social media site to reach a million users [1]. From 2004 to 2020, the number of users using the social platform is somewhere in billions. For example, TikTok, a social networking service for video sharing started around 2016, had almost half a billion users by 2019. Take another example, as of April 2020, more than 60 million songs, 2.2 million apps, 25,000 TV shows, and 65,000 films are available in iTunes [2]. This rise of social media, multimedia, and the Internet of Things (IoT) has accounted for a sharp and continuous increase in the volume of the data either in the structured or unstructured format. The data are created at a rate never witnessed before [3], referred to as big data has emerged as a well-recognized trend across the globe. From different sections of the society, like industry and government, big data are evoking consideration because of three reasons:

1. The volume of data is enormous
2. The data that have been generated cannot be fit into a regular relational storage model

DOI: 10.1201/9781003032328-3

3. Data are generated at a very high speed and need to be stored and processed at the same time

All the current sectors of organizations, such as health care, research and development, science and engineering, and the society itself are impacted by the growth of big data. More advancements in technology concerning data mining and storage have allowed organizations to accumulate and store large amounts of data for future prediction and analysis and has changed the way organizations use to make decisions [4]. With better internet speed and services, the rate at which the data is being created is staggering [5]. The main issue caused by this high-speed data generation is that the infrastructure available is insufficient to handle this data. This has ensured that the organizations and researchers find a framework/method by which this enormous data that are being created are not only stored but processed at the same time.

One of the significant shifts in information and communications technology has been the development of cloud computing. According to NIST [6], "cloud computing is used for providing on-demand computing resources which are configured, rapidly provisioned and released as per the requirement of users with minimal user interference". Users and organizations are able to perform complex and large-scale processing of tasks with resources being provisioned from the cloud. Parallel processing shared, virtualized, and scalable resources are some of the advantages that cloud computing offers for processing big data to organizations. Cloud computing provides sharable and virtualized resources that can be scaled up/down as per the requirement to organizations and individuals due to which the cost of procuring and managing the resources is minimized and organizations and users are able to maximize their profit [7]. These advantages have ensured that a large variety of applications that utilize infrastructure from the cloud are being developed rapidly. Such applications are also acting as rich sources for generating a large volume of data. Public cloud providers like Amazon Web Services are providing their users with Hadoop clusters for managing big data in cloud computing by providing a highly scalable and elastic environment [8].

Cloud computing can provide scalable resources to users due to virtualization. Virtualization in cloud computing acts as a base technology that helps the cloud service providers to provide resources to the user as per the demand. Virtualization allows the service providers to provide multiple machines to users running over a single physical machine and thus increasing the resource utilization. Resources to big data applications over the cloud are managed by the help of virtualization.

3.2   DEFINITION AND CHARACTERISTICS OF BIG DATA

Big data are used to define the enormous volume of data that cannot be stored, processed, or analyzed using traditional database technologies. These data or data sets are too large and complex for the existing traditional database. The idea of big data is vague and includes extensive procedures to recognize and make an interpretation of the information into new bits of knowledge. Although the term big data is available from the last century; however, its true meaning has been recognized after social media has become popular. Thus, one can say that the term is relatively new to the IT industry and organizations. However, there are several instances where the research-ers have used and utilized the term in their literature.

The authors in [9] defined a large volume of scientific data as big data required for visualization. Several authors have defined big data in different ways. One of the ear-liest was given by Gartner in 2001 [1]. The Gartner definition has no keyword as "big data"; however, he defines the term with three Vs: volume, velocity, and variety. He has discussed the increasing rate and size of data. This definition given by Gartner is later adopted by various agencies and authors such as NIST [10], Gartner himself in 2012 [11], and later IBM [12], and others include a fourth V to original three Vs: veracity. The authors in [13] explained the term big data as "the volume of data that is beyond the current technology's capability to efficiently store, manage, and process". The authors in [14] and [15] used Gartner's classification of 2001 of three Vs: volume, variety, and velocity to define big data (Figure 3.1).

So based on the discussion, the term big data can be defined as a high volume of a variety of data that are generated at a rapid speed, and the traditional database system is not fully functional to store, process, and analyze that in real time. Let us look into the three Vs that are defined by various authors in their work.

1. Volume is the measure of information created from a variety of sources that keep on growing. The major benefit of such a large collection of information is that it helps in decision making by identifying hidden patterns with the help of data analytics. Having more parameters let us say 200 to forecast weather will be able to predict weather better as compared to forecasting with 4–6 parameters. The volume in big data refers to the size of zettabytes (ZB – 10,007 bytes) or yottabyte (YB – 10,008 bytes). Thus, it becomes a huge challenge for the current infrastructure to store this amount of data. Most companies like to put their old data in archives or logs form, i.e., in an offline mode. But the disadvantage of this is that the data are not available

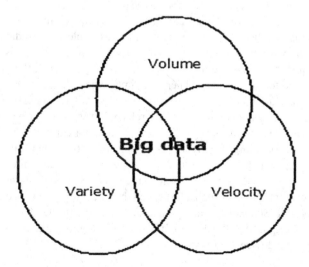

FIGURE 3.1   Three Vs of big data.

for processing. Thus, it requires a scalable and distributed storage that is being offered by cloud as cloud storage either in the form of object, file, or block storage.

Assuming that the enormous volume of the data cannot be processed by the traditional database, the options left for processing are either breaking the data into chunks for massively parallel processing frameworks such as Apache Hadoop or database like Greenplum. Using the data warehouse or database evolves the predefined schemas to be entered into the system, given the other Vs – variety: this is again not feasible for big data. Apache Hadoop does not place any such condition on the structure of data and can process it without a predefined structure. For storing data, the Hadoop framework used its own distributed file system known as Hadoop Distributed File System or HDFS. When data are required by any node, HDFS provides that data to the node. A typical Hadoop framework has three steps for storing data [16]:

- Loading data into HDFS,
- Performing operation of MapReduce, and
- Fetching results from HDFS.

This process is a batch operation by nature and is suited for computing tasks that are analytical or non-interactive.

2. Velocity refers to the rate at which data are transferred from users. The contents of information continually change as a result of the retention of integral information assortments, presentation of recently filed information or inheritance assortments, and gushed information coming up from various sources [9]. The velocity of data also follows a trend similar to that of volume. With more smart and handheld devices present in the market, and with better connectivity, the data are being generated at a rate never witnessed before. The financial institutes and the online stores are some of the few examples that are not only storing the data arriving at a fast pace but are also processing it at the same time. Processing data arriving at a high velocity is a huge challenge for organizations. For example, storing data for later processing while it is streamed at a high speed. The important issue is the time required to complete the feedback loop, i.e., processing the data that is being stored to take decision. There are times where the users do not like to wait for the Hadoop framework to complete a submit job. Such scenarios for such fast-moving data in industries are referred to as either "streaming data" or "complex event processing." There are two major causes for considering streaming processing. Firstly when storing the entire input data which is coming at a fast speed, some level of analysis is required while storing as the data are streaming in. For example, the Large Hadron Collider (LHC) at CERN generates data at such a fast speed that the scientist has to discard the majority of it. Secondly, to an application where the immediate response to the data is required as the data stream in. For example, while playing games at the mobile device requires this scenario [17].

The velocity of a system is not only about input but also about output too. The more feedback system is rigid; the greater will be its competitive advantage. The results obtained from this feedback may be directly fed into a product like the recommendation given by Amazon to purchase a product. This speed has ensured those key-value databases are developed for fast retrieval of information. Such databases are known as NoSQL databases and are used where the traditional database fails [18].

3. Variety refers to the various types of data collected over a large range of devices such as smartphones, sensors, or social networks. The collected data may be in a structured or unstructured format and can be comprised of logs, video, image, text, and audio. The data generated by mobile devices such as text messages, blogs, and gaming are unstructured in format. Users browsing on the Internet generate extremely varied data that are classified as structured or unstructured data [19]. With big data, the data will rarely be in a ready format that can be easily fit into a relational model and processed easily. Since the sources of the big data are too diverse, the data generated from them are also diverse. Even in computer to computer communication, different browsers send the data in a different format having information as per the language, software version, and communication protocol user uses. If the user tries to manage and process this data, it is natural that error and inconsistency will be there [20].

Usually, big data processing takes unstructured data and tries to extract meaning from it can be used by humans or an application. As you move the data from the source to the processed application, there is a chance that the information may be lost. And when you try to resolve this, you discard some pieces of information. The information that is being discarded may contain valuable information and here the principle of big data comes into the picture: keep everything when you can [21].

The different variety of big data can be classified into five different aspects depending on the sources from where they are collected, it's processing, and storage mechanism. This classification becomes important as in the cloud there is large scaling of data and segregation can be made based on classification. The different classifications of big data are based on:

- Data source
- Content format
- Data stores
- Data staging
- Data processing

Each of the classifications has its individual characteristics and intricacies. The different sources in each classification are depicted in Figure 3.2.

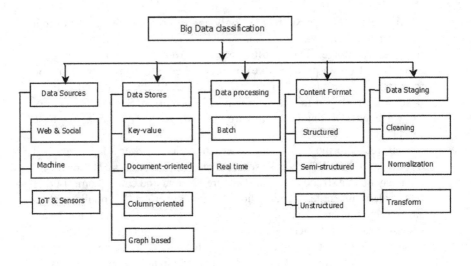

FIGURE 3.2 Big data classification.

3.3 CLOUD COMPUTING

Cloud computing is on-demand availability of virtualized resources such as computing power and storage to end-users without the direct involvement of service providers over the Internet. Cloud computing is one of the fastest-growing technologies that has impacted the traditional business and IT industry. Cloud computing provides its user with on-demand reliable virtualized hardware and software through an application programming interface (API) via the Internet [22]. The user needs to only pay for the resources as per its use. Cloud computing has become a powerful tool for executing a large and complex high-performance application that requires extensive infrastructure. Cloud computing has been adopted by various organizations and individuals for storing, processing, and analyzing large volumes of data that are growing every day [23]. Considering this the cloud service providers have also started integrating framework required for parallel processing of data to assist users for processing a high volume of data [24]. Cloud computing allows the organization to have total control over the virtualized environment and to have their focus on business issues, rather than thinking about the deployment and availability of infrastructure and resources [25].

Cloud computing provides its user resources and application as services, a delivery model that consists of Software as a Service (SaaS), Platform as a Service (PaaS), and Infrastructure as a Service (IaaS). Along with new services like Workflow as a Service (WaaS) [26], Model as a Service (MaaS) [27], and Data as a Service (DaaS), cloud computing is presenting the organizations and researchers the prospect of working on anything as a service (XaaS) [28].

Software as a Service or SaaS refers to applications running on cloud infrastructure that can be accessed by users as a service via the Internet [29]. Example of SaaS includes Google Docs and Gmail. Platform as a Service or PaaS provides a platform to users for developing, running, and managing applications. Example of PaaS

Google's Apps Engine, Salesforce.com Infrastructure as a Service or IaaS refers to the underlying virtualized hardware running over the physical infrastructure at the location of the service provider [30].

The biggest advantage of using cloud computing is the unlimited storage that the users can get. With the new handheld devices coming to the market are responsible for generating the bulk of big data today, these devices come with limited storage. The users can push the data to the cloud, without worrying about the limited storage that the device has. The cloud service provider has started providing the users with the tools not only to store but also to analyze and process the data that is coming at a high speed and volume.

Big data and cloud computing are integrated closely. The users can use commodity computing using big data and are able to process multiple databases spread across large geographical locations using multiple queries on time. Cloud computing helps the user by providing the underlying architecture required for processing of the big data through Hadoop. Using the algorithms for parallel and distributed environments in a cluster, large distributed data sets are processed that are stored in the cloud. The key goal of data visualization is to examine analytical results obtained after processing of data in the form of graphs for making decisions (Figure 3.3) [31].

Cloud computing offers big data and large distributed remote storage for storing data. Using virtualized technologies, cloud computing can run applications that require big data for their execution. Thus, cloud computing using virtualization technology is able to provide amenities required by big data for computation and processing. As discussed above, the cloud provides the resources to its user as services, i.e., IaaS, PaaS, and SaaS. Let us look at how these services are used to deliver big data services to users.

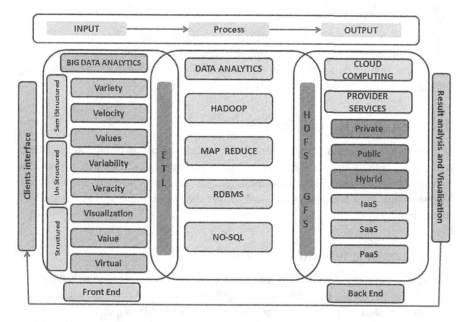

FIGURE 3.3   Relationship model of big data and cloud computing.

3.3.1 IaaS in Public Cloud

Using the IaaS service of cloud, users are able to use unlimited computing power and storage that is essential for big data. The IaaS service is a cost-effective solution for enterprises using big data for analysis as the cloud service provider bears the entire operating cost of managing the physical hardware [32].

3.3.2 PaaS in Private Cloud

Vendors offering PaaS as service integrate big data technologies into PaaS. Therefore, they remove the requirement of managing complexities concerning the software and hardware elements while executing terabytes of data.

3.3.3 SaaS in Hybrid Cloud

SaaS vendors provide the users with tools to conduct the analysis required to analyze the social media data.

Thus, cloud computing also acts as a service model for big data, integrating the three service models for storing, processing, and analyzing big data. Thus, the cloud empowers the "As-a-Service" design by abstracting the difficulties and intricacy by an adaptable and versatile self-administration application. The prerequisite for big data is similar where dispersed handling of enormous information is preoccupied with the clients. Using big data analytics in the cloud offers multiple advantages as follows.

3.3.4 Improved Analysis

Big data analysis is giving better outcomes as it has become better due to advancements in the technology of cloud computing. Thus, organizations have started to prefer to carry out big data analysis in the cloud. Besides, cloud assists in coordinating information from various sources [33].

3.3.5 Simplified Infrastructure

With data coming in huge volumes at differing rates, analysis of big data is a huge demanding activity on infrastructure and the traditional infrastructures are not able to handle this volume and speed. As the cloud offers flexible and portable on-demand infrastructure, which the users can scale as per the requirements, it becomes effortless to handle workloads [34].

3.3.6 Lowering the Cost

Combining both cloud computing and big data helps an organization to reduce cost by reducing the ownership while delivering the services. The originations can minimize both capital cost (CAPEX) and operational cost (CAPEX), as the cloud offers the pay per use model to them. Since the most popular framework for big data is

Hadoop which is distributed under Apache license allows organization to save cost in millions required to assemble and purchase. Consequently, both cloud and big data are helping the organization to bring the expense down for business purposes [35].

3.3.7 SECURITY AND PRIVACY

The two significant concerns while managing enterprise data are information security and privacy. As the cloud is open and has limited control on clients when an application is deployed on the cloud, the security of the application turns into an essential concern. Big data is an open source; thus, it utilizes numerous solutions that are offered by third party. Nowadays, the Private Cloud Solutions are also elastic and scalable and also use Scalable Distributed Processing [36].

The data in the cloud are stored and processed in a central data center, i.e., cloud storage. The cloud service providers sign a service-level agreement with its end users for providing the services and to increase the level of trust between them. If required by the client, the service provider may also offer an additional level of security control to users' data. Thus, the security issues of big data in cloud computing include:

1. Securing data from complex and new threats
2. Management of storage and data by the service provider

3.4 BIG DATA IN CLOUD COMPUTING

Cloud computing offers a large pool of virtualized resources to store, process, and analyze big data. Let us look at how these are supported in cloud computing as such.

3.4.1 BIG DATA STORAGE IN CLOUD COMPUTING

With an enormous volume of data, storage is a huge challenge for big data. Cloud computing with its large pool of virtualized resources can provide big data with unlimited elastically expandable storage to support the volume of big data. To provide storage to this enormous volume of data, the cloud provides service in the form of block storage as Data Storage as a Service (DSaaS). The service can add external storage to store data in the form of blocks. This allows the cloud to add more storage volume without physically loading hard drives. With an unlimited pool of virtualized resources, the users using cloud computing can satisfy their requirements of storage dynamically for storing data that has huge volume and velocity. The users can also use external data storage as backup storage to keep the backup data so that nothing is lost and data are safe at back-end servers and can be restored easily. Moreover, since the physical location is unknown, the security of data is also provided [37].

3.4.2 BIG DATA PROCESSING IN CLOUD COMPUTING

To process huge volumes of data, dedicated resources such as CPUs and high-density disks and RAMs are required [38]. The resources such as computing, network, and storage are provided by cloud computing in demand to its user. Usually, the power consumed by a resource is about two-thirds to that of an active resource [39]. To improve system stabilities, for a speedy recovery, separate virtual structures are created for big data systems to manage various backup images and file systems. The capability of the environment to replicate the nodes in virtual machine clusters increases the utilization of resource pools for supporting big data analytics. With a large virtualized resource pool providing the basic requirement for storing big data, data processing utilizes the advantage for quick acquisition and relocation of data. Even though to reduce the bottleneck for processing big data, cloud computing is an exceptional choice, however, several issues like options available for cloud storage for storing the input and output and the bandwidth requirement as per the application makes it less appropriate [40].

3.4.3   BIG DATA ANALYTICS IN CLOUD COMPUTING

Apache Hadoop, the most recognized tool for big data analytics, is normally installed on a cluster of physical machines, which results in the wastage of computing resources as there is hardware redundancy. Cloud computing can provide these clusters as virtualized clusters using virtualization technology, and thus the analytical platforms are migrated from physical clusters to virtual, optimizing the utilization of computing resources [41].

Using the help of tools such as autoscaling and load balancing, the analytical frameworks can be deployed for scalable and on-demand analytical platforms and thus computing resources wastage can be minimized. Scalability in distributed architecture along with parallel execution of jobs can be achieved through autoscaling. When a program that has parallel execution is enabled, autoscaling helps in fetching resources required for execution. If autoscaling is unavailable, then public cloud service providers will not be able to provide automatic scalability [39]. For even distribution of workload in virtual cluster, the load balancer is used by the cloud service providers. If higher computing configurations are required by analytics, the balancer then triggers autoscaling. For a better resource allocation, load balancer can be optimized and can also act as a distributor for managing virtual network traffic (Table 3.1).

3.5   CONCLUSION

The volume of data is enormous and is increasing at a rapid rate. Along with this volume, the variety of data is also increasing. With the increase in the number of mobile devices and other devices that are connected to the Internet like sensors, the rate of generation and growth of data have also increased. These data are helping the organization to gain new insights that too in real-time and helping them to create new opportunities. With the advancement in the Internet and Information and communications technology, the growth of cloud computing has given organizations tools for storing, processing, and analyzing data in real time with on-demand delivery. In this chapter, the relationship between big data in cloud computing is discussed. For data processing, cloud computing edges over traditional computing tools. The customizable on-demand resources help the organizations to save computing resources with higher flexibility and security to data storage.

TABLE 3.1

The Different Service Providers That Provide Different Solutions for Big Data

	Amazon Web Services	Google Cloud Platform	Azure Cloud	IBM Cloud
Big data storage	S3	Google cloud services	Azure	IBM Elastic Storage
MapReduce	Elastic MapReduce	AppEngine	Hadoop on Azure	Hadoop on IBM
Big data analytics	Elastic MapReduce	BigQuery	Hadoop on Azure	IBM Analytics Engine
Relational database	MySQL or Oracle	Cloud SQL	SQL Azure	Db2
NoSQL database	DynamoDB	AppEngine Datastore	Table storage	IBM Cloudant
Streaming processing	Nothing prepackaged	Search API	Streaminsight	IBM Streaming Analytics
Machine learning	Hadoop + Mahout	Prediction API	Hadoop + Mahout	IBM Watson
Data import	Network	Network	Network	Network

REFERENCES

1. L. Douglas. 3D data management: Controlling data volume, velocity and variety. Gartner. Retrieved, 6, 2001.
2. https://www.couponxoo.com/apple-music-song-promotion, accessed on 01.08.2020.
3. R.L. Villars, C.W. Olofson, M. Eastwood. Big data: What it is and why you should care, White Paper, IDC, 2011, MA.
4. R. Cumbley, P. Church. Is big data creepy? *Comput. Law Secur. Rev.* 29 (2013) 601–609.
5. S. Kaisler, F. Armour, J.A. Espinosa, W. Money. Big data: Issues and challenges moving forward, system sciences (HICSS), *Proceedings of the 46th Hawaii International Conference* on, IEEE, 2013, pp. 995–1004.
6. Mell, Peter, T. Grance. The NIST definition of cloud computing, 2011.
7. L. Chih-Wei, H. Chih-Ming, C. Chih-Hung, Y. Chao-Tung, An improvement to data service in cloud computing with content sensitive transaction analysis and adaptation, *Computer Software and Applications Conference Workshops (COMPSACW), 2013 IEEE 37th Annual*, 2013, pp. 463–468.
8. L. Chang, R. Ranjan, Z. Xuyun, Y. Chi, D. Georgakopoulos, C. Jinjun. Public auditing for big data storage in cloud computing – A survey, Computational Science and Engineering (CSE), *2013 IEEE 16th International Conference on*, 2013, pp. 1128–1135.
9. M. Cox, D. Ellsworth. Managing big data for scientific visualization, ACM Siggraph, MRJ/NASA Ames Research Center, 1997.
10. NIST Big Data Working Group (NBD-WG). http://bigdatawg.nist.gov/home.php.
11. M.A. Beyer, D. Laney. *The Importance of Big Data: A Definition*. Stanford, CT: Gartner, 2012.
12. IBM. What is big data? – Bringing big data to the enterprise. http://www-01.ibm.com/software/data/bigdata/, July 2013.

13. J. Manyika, M. Chui, B. Brown, J. Bughin, R. Dobbs, C. Roxburgh, A.H. Byers. Big data: The next frontier for innovation, competition, and productivity, 2011.
14. P. Zikopoulos, K. Parasuraman, T. Deutsch, J. Giles, D. Corrigan. *Harness the Power of Big Data The IBM Big Data Platform*. New York: McGraw Hill Professional, 2012.
15. J.J. Berman. Introduction, in: Andrea Dierna (Ed.), *Principles of big data: Preparing. Sharing, and Analyzing Complex Information* (1st ed.). Boston, MA: Morgan Kaufmann, 2013, pp. xix–xxvi.
16. https://www.forbes.com/sites/oreillymedia/ 2012/01/19/volume-velocity-variety-what-you-need-to-know-about-big-data/?cv=1#e616db21b6d2.
17. S. Begam, J. Vimala, G. Selvachandran, T.T. Ngan, R. Sharma. Similarity measure of lattice ordered multi-fuzzy soft sets based on set theoretic approach and its application in decision making. *Mathematics* 8 (2020) 1255.
18. T. Vo, R. Sharma, R. Kumar, L.H. Son, B.T. Pham, B.D. Tien, I. Priyadarshini, M. Sarkar, T. Le. Crime rate detection using social media of different crime locations and twitter part-of-speech tagger with brown clustering, 2020, pp. 4287–4299.
19. D.E. O'Leary. Artificial intelligence and big data, *IEEE Intell. Syst.* 28 (2013) 96–99.
20. P.T. Nguyen, D.H. Ha, M. Avand, A. Jaafari, H.D. Nguyen, N. Al-Ansari, T. Van Phong, R. Sharma, R. Kumar, H.V. Le, L.S. Ho, I. Prakash, B.T. Pham. Soft computing ensemble models based on logistic regression for groundwater potential mapping, *Appl. Sci.* 10 (2020) 2469.
21. S. Jha, et al. Deep learning approach for software maintainability metrics prediction, *IEEE Access* 7 (2019) 61840–61855.
22. M. Armbrust, A. Fox, R. Griffith, A.D. Joseph, R. Katz, A. Konwinski, G. Lee, D. Patterson, A. Rabkin, I. Stoica, M. Zaharia. A view of cloud computing, *Commun. ACM* 53 (2010) 50–58.
23. L. Huan. Big data drives cloud adoption in enterprise, *IEEE Internet Comput.* 17 (2013) 68–71.
24. S. Pandey, S. Nepal. Cloud computing and scientific applications — Big data, *Scalable Anal. Beyond, Futur. Gener. Comput. Syst.* 29 (2013) 1774–1776.
25. R. Sharma, R. Kumar, D.K. Sharma, L.H. Son, I. Priyadarshini, B.T. Pham, D.T. Bui, S. Rai. Inferring air pollution from air quality index by different geographical areas: Case study in India, *Air Qual. Atmos. Health* 12 (2019) 1347–1357.
26. M. Krämer, I. Senner. A modular software architecture for processing of big geospatial data in the cloud. *Comput. Graphics* 49 (2015) 69–81.
27. C. Li, et al. Trust evaluation model of cloud manufacturing service platform. *Int. J. Adv. Manuf. Technol.* 75(1) (2014a) 489–450.
28. C. Yang, et al. Utilizing cloud computing to address big geospatial data challenges. *Comp. EnvironUrban Syst.* 61 (2017a) 120–128.
29. A. O'Driscoll, J. Daugelaite, R.D. Sleator. 'Big data', Hadoop and cloud computing in genomics. *J. Biomed. Inform.* 46 (2013) 774–781.
30. R. Sharma, R. Kumar, P.K. Singh, M.S. Raboaca, R.-A. Felseghi. A systematic study on the analysis of the emission of CO, CO_2 and HC for four-wheelers and its impact on the sustainable ecosystem, *Sustainability* 12 (2020) 6707.
31. S. Sharma, et al. Global forecasting confirmed and fatal cases of COVID-19 outbreak using autoregressive integrated moving average model, *Front. Public Health* (2020). https://doi.org/10.3389/fpubh.2020.580327.
32. P. Malik, et al. Industrial Internet of Things and its applications in industry 4.0: State-of the art, *Comp. Commun.* 166 (2021) 125–139, Elsevier.
33. Analysis of water pollution using different physico-chemical parameters: A study of Yamuna river, *Front. Environ. Sci.* https://doi.org/10.3389/fenvs.2020.581591.
34. D. Dansana, et al. Using susceptible-exposed-infectious-recovered model to forecast coronavirus outbreak, *Comp. Mater. Continua* 67(2) (2021) 1595–1612.

35. M.T. Vo, A.H. Vo, T. Nguyen, R. Sharma, T. Le. Dealing with the class imbalance problem in the detection of fake job descriptions, *Comp. Mater. Continua* 68(1) (2021) 521–535.
36. S. Khoudali, K. Benzidane, A. Sekkaki, M. Bouchoum. Toward an elastic, scalable and distributed monitoring architecture for cloud infrastructures. *2014 International Conference on Next Generation Networks and Services (NGNS)* (pp. 132–138), IEEE, 28 May 2014.
37. K. Mayama, et al. Proposal of object management system for applying to existing object storage furniture, *2011 IEEE/SICE International Symposium on System Integration (SII)*, IEEE, Kyoto, Japan, 20–22 Dec 2011.
38. C. Yang, et al., Utilizing cloud computing to address big geospatial data challenges, *Comp. EnvironUrban Syst.* 61 (2017a) 120–128.
39. A.D. Josep, R. Katz, A. Konwinski, L.E.E. Gunho, D. Patterson, A. Rabkin. A view of cloud computing, *Commun. ACM* 53 (2010) 4.
40. S. Sachan, R. Sharma, A. Sehgal. Energy efficient scheme for better connectivity in sustainable mobile wireless sensor networks, *Sustainable Computing: Informatics and Systems* 30 (2021) 100504.
41. S. Ghanem, et al. Lane detection under artificial colored light in tunnels and on highways: An IoT-based framework for smart city infrastructure, *Complex & Intelligent Systems* (2021). https://doi.org/10.1007/s40747-021-00381-2.

4 Effect of the Measurement on Big Data Analytics
An Evolutive Perspective with Business Intelligence

M. J. Diván and M. Sánchez-Reynoso
National University of La Pampa

CONTENTS

4.1 THE ROLE OF THE MEASUREMENT IN THE DECISION-MAKING

The current economy, globalization, and the convergent technology have produced an unprecedented change in the way in which people live and work (Brous et al. 2020). Regional economies and their relationships with extra-zone partners produce different kinds of variations in the per-capita income of countries (Zhao and Serieux 2020). However, companies must continue working even when the environment is constantly changing and affecting them directly or indirectly (Laukkanen and Tura 2020).

The decision process incorporates different stages and analysis before a decision is made (Chiheb et al. 2019). Briefly, a comprehension about the situation and the fact requiring a decision jointly with the specification of assumptions and hypotheses are carried out. After, an analysis of the context where the decision could impact is carried out. Next, alternatives and their feasibilities are specified from the technical, operational, and economic perspectives. Once the alternatives are available, the

DOI: 10.1201/9781003032328-4

decision criteria are established to produce a scoring that allows ordering from the most to the east beneficial alternative. Thus, the most beneficial alternative is chosen from the list, keeping in mind the rest ones in case the first one is no longer available (independently the reason). Different scenarios modeling about the context could be incorporated to fit the scoring according to each one (Ramalho et al. 2019).

In Chiheb et al. (2019), interaction between the "Decision model" and the "Data collection and preparation" into the "Intelligence" phase is proposed. However, it is worthy to mention that decision-makers use the information for supporting their decisions and no data; this constitutes the origin of the data mining and the evolution of different concepts of the abstraction capacity in reinforcement learning as a way to model the knowledge (Ho et al. 2019). Data mining could be defined as an application of different techniques on huge data volume to discover non-trivial, useful, interesting, and applicable information, for example, for the bag-of-concepts representation associated with document classification (Li et al. 2020).

Meaningful differentiation is incorporated among data, information, and knowledge. The data are associated with some fact that is captured and stored. Nothing is said about the pertinence or quality of such data. However, and as a difference, the information is a data that concurrently is consistent, opportune, truth, and interesting according to the decision-maker perspective (Artiga 2020). Thus, the information is a datum, but a datum is not always information. The difference between information and knowledge turns around the results (Anila Glory et al. 2020). That is to say, knowledge requires that the application or use of given information produces quantifiable results. For example, it is important to know what the fever is but also to know the way to decrease the corporal temperature in a person. The possibility to recognize the fever is information, but how to decrease the fever using a given method is knowledge because it is possible to quantify the time and resources employed to reach the aim.

The measurement is a process where a comparison between a given measure unit and an object or concept is performed to produce a quantitative value in terms of the unit (Krechmer 2018). For example, a kilowatt is a unit of power that represents the consumption rate in function of the time. When the energy consumption needs to be measured on an object (e.g. a data server), the obtained magnitude represents how many times the chosen unit is contained in it.

A decision-maker needs to get the last known data from the concept under analysis (i.e. a business process, an autonomous vehicle, etc.) to make a decision based on data as updated as possible. This represents a key asset in the businesses where a delay could represent an advantage for a competitor. Thus, the relationship between the decision-making and measurement process is worthy to analyze because the experimental design provides data useful, truth (with statistical support), consistent (it is based on a given method), and interesting to those who need to use them.

The Internet of Things (IoT) has allowed to reach cheap, available, and easily accessible devices, fostering complex and interesting data collection strategies to implement and automatize measurement process, at the time in which some challenges related to energy consumption, battery autonomy, and environment care, among others taken an special consideration (Lemoine et al. 2020). Different applications in Smart Cities (Fatimah et al. 2020), Smart Farms (Ramli et al. 2020), and

Autonomous vehicles (Sachdev et al. 2019), among other areas, have emerged as study cases of many kinds of business models, describing opportunities and risks (Lee 2019).

Thus, the possibility to easily generate data coming from sensors, the World Wide Web, among other data sources impacted on the volume of data, producing a big data environment with two contexts differentiated: on the one hand, a context in which a huge data volume is processed and analyzed, following a batch data processing schema (e.g. meteorological repositories); on the other hand, a context in which the data are processed and analyzed as soon as they arrive, fostering real-time monitoring (e.g. the stock market monitoring). Both contexts have different kinds of impact on the environment; that is to say, while the batch data processing on huge data repositories has a high energy consumption, the IoT devices in real-time data processing use batteries as a primary way to keep its autonomy.

In this chapter, the effect of the measurement process on the big data analytics is contrasted with the business intelligence perspective, highlighting similarities, differences, and the natural evolution of a set of terms and concepts. Also, the incidence of the IoT and data streaming on big data analytics is compared to the batch data processing strategy. Thus, the relationships among measurement, scenarios, and entities under monitoring are analyzed under the light of the analytics, considering their effects on the data quality.

Thereby, once the role of the measurement on the decision-making and its perspective about big data are introduced, the rest of the chapter is organized as follows. Section 4.2 describes the emergence of business intelligence and its impact on business models. Section 4.3 introduces the impact of data quality on big data analytics and the perspective of the real-time data processing. Section 4.4 outlines the idea of scenarios and entity states associated with the measurement process jointly with its impact on the decision-making process. Section 4.5 presents aspects to be monitored in the evolution of scenarios or states and their impact on big data analytics. Section 4.6 proposes an integrated view about the information-driven decision-making, data quality, batch and online data processing and its impact on the big data analytics. Finally, some conclusions are highlighted, and future works are described.

4.2 THE EMERGENCE OF BUSINESS INTELLIGENCE: A BUSINESS PERSPECTIVE

Decision support systems could be defined such as those systems oriented to support the decision-making process based on information and knowledge obtained and organized in such a way that allows providing recommendations as close to a situation as possible. It emerged in the early 1960s, evolving up to reach a certain maturity at the end of 1980s (Keenan and Jankowski 2019). At that moment, the data volume generation and interchanging had taken particular importance mainly influenced by the World Wide Web which born in 1989 and the globalization of communications (Martins et al. 2015). Thus, before 1989, companies were able to manage a certain volume of data and they were working on certain kinds of business strategies mainly focused on personal contact. However, the Internet's birth opened the doors in very different ways to make business based on a global and massive electronic market based on technology, where the end-user becomes a proactive analyst of each offer having the possibility to compare options easily (Song and Zahedi 2006; Sterling and LeRouge 2019).

The beginning of the 1990s was characterized by different kinds of report strategies, while the data volume started to grow exponentially due to the Internet. Early, the companies realized the huge data volume that they needed to process to reach an objective, and how the difference between the concepts of data and information had been highlighted. Thus, strategies oriented to pre-process data from the Online Transactional Processing (OLTP) systems, among others, for integrating, solving inconsistencies, and homogenizing transactional data through the Operational Data Stores (ODS) emerged. However, due to the managed data volume, decision-makers started to demand different levels of aggregation to improve the data analysis, trying to reach the concepts expressed in a business language to interpret an economic fact from different points of view. Thus, the concepts of Data Warehouse (DW) and Online Analytical Processing (OLAP) emerged as a potential solution with the idea of avoiding relying on IT Staff (Chaudhuri et al. 2011; Kurnia and Suharjito 2018).

This produces an explicit change of the kind of query strategy according to the kind of underlying model used to represent the data. On the one hand, the OLAP technology used the multidimensional data modeling that allows ad-hoc queries under a business language from a limited number of users and expecting a huge data volume as an answer. On the other hand, the OLTP technology was based on the relational data model, fostering the data transactions from a huge number of users, structured queries, with short answers through the Structured Query Language (SQL) which was not friendly for business. Both technologies kept complementary between them for business because they are understood as oriented to satisfy different business necessities. OLTP was oriented to an operative level, while OLAP was oriented to decision-makers at least in tactical organizational levels (Plattner 2009).

Earliest in 2000, the volume of data had increased considerably no just in terms of volume but also in terms of heterogeneity, incorporating an additional complexity to the data analysis. Managers start to need new tools to discover valuable, interesting, consistent, and applicable data from the data's sea, in that context the data mining emerges as an alternative. What manager needs indeed is to justify the maintenance cost of a give data processing infrastructure, finding new knowledge that allows them to make a difference in the businesses. The batch data processing was the main approach to develop the data mining functions based on the Knowledge Discovery in Databases (KDD) process which added a new perspective to Business Intelligence (Safhi et al. 2019). However, as a touch of color, in 1999 the concept of Internet of Things was introduced by Kevin Ashton associated with the RFID technology, cultivating the field for the data streaming paradigm (Chin et al. 2019).

In 2005 emerged the concept of big data coined by Roger Mougalas to address the challenge of the huge data volume, which required a different processing approach oriented to a cluster-based distributed batch data processing (Yaqoob et al. 2016). In the same year, it was born Hadoop as an Apache Foundation project based on the MapReduce programming model, opening the doors to develop open-source software for reliable, scalable, and distributed computing. The original data mining was potentiated through projects such as Apache Mahout which introduced the data mining functionality as a part of the Hadoop ecosystem (Landset et al. 2015).

The evolution of IoT and the introduction of the data streaming model during the first decade of 2000 incorporated a new perspective in big data related to the processing style. Traditionally, the data processing was based on batch processing (distributed or not), while the IoT and data streaming talked about data coming in real time that need to be processed as soon as they arrive using just the current resources. When the data could not be processed for any reason, it was discarded because always new data would be available. This produced a great turn due to the algorithms needed to be incremental; for example, a decision tree could be trained based on a huge data set in big data, but in data streaming, the classifier needed to be incremental and to learn from each datum while it was processed (Sahal et al. 2020). These kinds of approaches quickly called the attention of different businesses which required online monitoring, for example, stock markets, telecommunications, advertising, and logistic, among others (Asghari et al. 2019).

Currently, any person could develop a wireless sensor network using IoT to implement different data collection strategies for supporting the decision-making process due to the cheap, available, and accessible to such hardware. However, it is important to highlight that this is complementary to the batch data processing, and it is not a replacement; each one has a given set of requirements in which they could contribute.

On the one hand, business intelligence is oriented to provide a business-friendly environment avoiding the dependence on the IT staff to support the decision-making process based on historical data. On the other hand, data analytics mainly focuses on studying, analyzing, and discovering different kinds of data relationships incorporating predictive abilities (be it online or not).

4.3 ABOUT THE DATA QUALITY AND BIG DATA ANALYTICS

The perspectives of big data analytics should be interpreted in the context. An analysis of the expected behavior is not isolated from an information need or context in which some situation needs to be analyzed or studied. Figure 4.1 introduces a contextualized perspective of big data analytics considering the information need (i.e. the objective), the business knowledge, and the application itself in addition to the modeling.

The information need determines the alignment of the big data analytics strategy. Thus, business knowledge is used along with the data pre-processing stage to solve the fundamental data problems. In this sense, it is possible to describe four fundamental data problems that are able to affect an analysis (Luo and Kareem 2020; Chen et al. 2020; Choudhury and Pal 2019; Gao et al. 2020):

- **Consistency**: It refers to the coherence and cohesion related to the data and the represented fact. That is to say, it analyzes the measure in which data represent a given concept, while the cohesion indicates how related is the data with their other meaningful neighbors. For example, the absence of linked data could identify a kind of risk associated with partial or total data isolation.

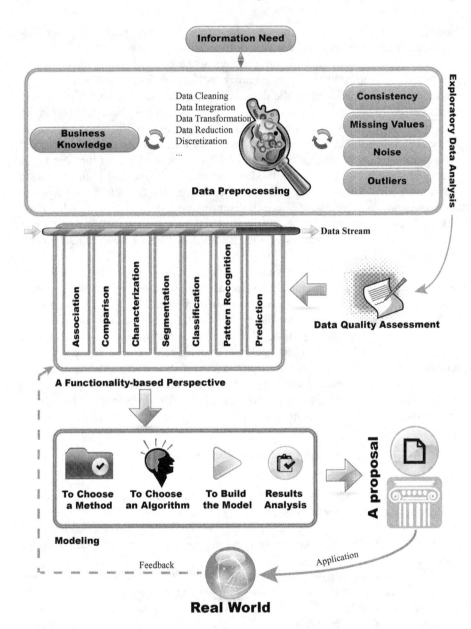

FIGURE 4.1   A contextualized perspective of big data analytics.

- **Missing Values**: It indicates the absence of a value for a given variable or attribute, but it does not represent a mistake itself. For example, the absence of the death date in the clinical story of a person indicates that such a person is alive, and it is not a data problem; the datum absence itself has an associated meaning. For that reason, the idea of completing missing values with some value could imply a bias.

- **Noise**: It represents an unexplainable variability due to random behavior, which implies the absence of causality. The last is very important to mention because the noise is random by nature; that is to say, when the data deviations have some cause that would imply that such deviations could be avoided, the randomness would not be present, and for that reason, the noise would be discarded.
- **Outliers**: It is associated with data that do not follow the expected general behavior for a given variable. Even when they are atypical values, they are not erroneous data. For example, an environmental thermometer could send values between 37°C and 38°C, and in a given moment, it informs a value of 45°C. That situation could be associated with the closeness of fire in relation to a sensor and not with a mistake or miscalibration itself.

To interpret and carry forward an adequate treatment of each fundamental data problem in context, business knowledge and the aim are essential to make decisions. Once the data set was treated, an evaluation of the data quality is recommended to quantify the level of adjustment of them based on requirements derived from the aim. It is possible to use conceptual models such as ISO 25012 or similar to analyze the level of quality of data (Merino et al. 2016). This aspect is particularly important and discriminates the real-time data processing from the batch data processing in big data analytics. That is to say, on the one hand, the batch data processing can analyze the data quality before carrying out the choice of a given analysis function. On the other hand, the real-time data processing associated with the data streaming context consumes the data such as it is without any previous modification.

Big data analytics could be analyzed from the functional perspective, as follows:

- **Association:** The aim is associated with the searching of different hiding relationships between data.
- **Comparison:** The focus is applied to the contrasting of two concepts, determining the level of similarity between them.
- **Characterization:** It pursues the discrete description of a given concept through their attributes.
- **Classification:** It tries to approach the behavior associated with an ordinal or categorical variable based on a set of attributes related to the concept. Thus, given a set of particular values for attributes, the model would approach the value for the target value with a given likelihood.
- **Pattern Recognition:** It is oriented to identify concurrent behaviors in the data to approach different kinds of joint occurrence.
- **Prediction:** In this case, the target variable is quantitative, so the idea is to reach a characteristic model integrated by a set of related attributes that allow approaching the quantitative value in a future time.
- **Segmentation:** It is focused on discriminating a given data set finding characteristic groups that maximize the variance between them, and at the same time, minimize the internal variability.

These big data analytics kinds do not imply the impossibility of interacting between them; that is to say, the use of segmentation jointly with characterization would be possible. The segmentation could obtain the groups on a data set, while the characterization would provide a way to explain each one.

Thus, once the functionality was chosen according to the level of maturity of data and the specified aim, the method and algorithm are chosen to reach a given model. The model is internally analyzed following different criteria of validation. Once the model is ready to be applied, it is effectively implemented in a real environment which will produce feedback. The feedback is provided iteratively to review the functionality/method/algorithm and its accordance with the aim.

Previously, the impact of data streaming on big data was mentioned, but here it can be visualized in terms of the data quality. As it is possible to appreciate in Figure 4.1, big data analytics based on batch data processing could address the treatment of data before applying any model. That is an advantageous situation because it is possible to provide a certain level of certainty and it could be quantified through a given level of data quality. However, such an advantage could be understood as a weakness in terms of the level of updating against the online data processing, where the model deals with the last known data for a given concept. Of course, such an advantage in the data streaming context has a cost, and it is that the data should be processed in the same way that they arrive, using the available resources in such a time. This opens the possibility of the presence of outliers, different kinds of inconsistencies, and noisy data, among other aspects that could affect the analysis or associated decisions.

4.4 THE IMPACT ON SCENARIOS AND ENTITY STATES ON THE MEASUREMENT

The measurement process consists of contrast a magnitude to a unit. The unit could be established by law or convention, but it is clearly defined before any comparison. Once the unit is defined, the measurement contrasts a given magnitude against the unit to detail how many times the unit is contained into the magnitude. For example, when it is said that a person has 1.83 m of height, indeed it is said that the "meter" unit is contained 1.83 times in the magnitude obtained using a given instrument. In this way, an instrument is said calibrated when it can contrast a measured unitary value (i.e. 1) against a given reference unit, considering a maximum tolerated error. In other words, it would like to know whether it is obtained one meter as a measurement, it is 1 m in terms of the international system or not (Krechmer 2016).

The *scale* represents the set of values that a variable could eventually assume. The kind of scales could be categorical, ordinal, interval, and rate. In the "categorical" scale, each value is understood as a category, a variable assumes only one category because the categories are mutually exclusive. In the "ordinal" scale, the categories incorporate a given order, being able to establish a hierarchy (e.g. Excellent, Very Good, Good, Regular, Bad). The "interval" scale is associated with numbers where the difference between them has a sense and where it is possible to establish a hierarchy or difference, but not a rate (e.g. the temperature expressed in degree Celsius). Finally, the "rate" scale consists of absolute numbers where their difference and rate have a meaning, being able to establish hierarchies, differences, and rates (e.g. number of items in the stock). The different scales give origin to the quantitative and qualitative variables that will then be used in different analyses. Thus, the

TABLE 4.1

Precision Versus Accuracy in the Measurement Process

	$t-4$	$t-3$	$t-2$	$t-1$	t
Ruler	1.97	1.98	1.97	1.97	1.97
Infrared device (ID)	2.01	2.00	1.99	2.00	2.01

measurement error could be interpreted as the difference between the measured or obtained value and the ideal and expected value according to the phenomena under analysis. This definition is useful to discriminate between accuracy and precision, initially saying that they are not synonymous. The precision refers to the possibility to provide the same value repeating the measurement over time in the same conditions; that is to say, the sum of the differences between the obtained values from the measured values from a certain attribute under measurement would tend to be zero. However, the accuracy refers to the possibility to approach the ideal value; that is to say, the sum of the differences between the obtained values and the ideal value representing the concept would tend to be zero. For example, suppose that it is known that the height of a door is exactly 2 m and to measure the height it will be used one ruler and one infrared device (Krechmer 2018). The door is measured five times using a ruler and an infrared device obtaining the values indicated in Table 4.1.

In terms of precision, the successive absolute differences would allow approaching a magnitude of the precision error after repetitive measures, as it is shown in Equations 4.1 and 4.2.

$$\text{Ruler}_{\text{precision}} = |1.97-1.98|+|1.98-1.97|+|1.97-1.97|+|1.97-1.97| = 0.02 \qquad (4.1)$$

$$\text{ID}_{\text{precision}} = |2.01-2.00|+|2.00-1.99|+|1.99-2.00|+|2.00-2.01| = 0.04 \qquad (4.2)$$

Following the example, the ruler would be more precise than the infrared device. However, it does not say anything about the approaching to the real value, but it refers to the minimization of the error through successive measurement carrying forward under the same conditions. For this example, it is known that the door's height is 2 m; for that reason Equations 4.3 and 4.4 refer to the accuracy in terms of how close to the ideal value are the measures.

$$\text{Ruler}_{\text{accuracy}} = |2.00-1.97|+|2.00-1.98|+|2.00-1.97|$$
$$+|2.00-1.97|+|2.00-1.97| = 0.14 \qquad (4.3)$$

$$\text{ID}_{\text{accuracy}} = |2.01-2.00|+|2.00-1.99|+|1.99-2.00|$$
$$+|2.00-2.00|+|2.00-2.01| = 0.04 \qquad (4.4)$$

The device closest to the ideal value is the infrared device, even when it has lower precision in this example than the ruler. It is important to highlight this aspect and its effect on the data both business intelligence and big data analytics because this has a great impact on the concept which the data represent.

Figure 4.2 describes a conceptual perspective related to the impact of the measurement jointly with scenarios and entity states on the data analysis. As it is possible to appreciate, the entity category under analysis represents a concept that needs to be monitored. The monitoring of each concept (e.g. a person) is characterized by a set of attributes (e.g. heartbeat). Analogously, the place or environment in which an entity category could have some influence on is described by a context. The context is characterized using a set of context properties that incorporates a discrete approach. Both attributes and context properties are quantified using metrics that contain the method, scale, unit, and instrument, among other aspects to get the quantitative value (i.e. the measure). Each measure is interpreted using decision criteria incorporated in indicators, but it is known that an entity could adapt its behavior according to its state and the current environment.

Thus, from the combination of attributes, a set of entity states could be identified according to the project requirements (e.g. resting, running, walking). In the same

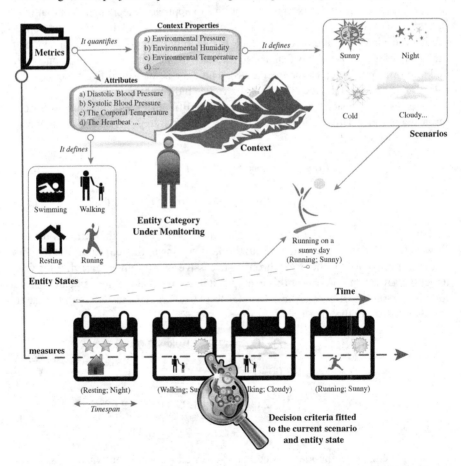

FIGURE 4.2   The impact of measurements, scenarios, and entity states on the data analysis.

way, the combination of context properties could be useful to define different scenarios to analyze the entity's behavior (e.g. cloudy, night, sunny). This allows suiting each decision criterion from the indicator depending on the scenario and entity state, producing a direct impact on the data analysis and the decision-making process. That is to say, think about the heartbeat, a value of 120 bpm (beat per minute) it would be fine while a person is running on a sunny day, but it would be risky if the person is resting. In other words, the same value could be interpreted differently depending on the current scenario and entity state.

Because each scenario is defined by a combination of context properties and their potential values, the current scenario could be approached by reading the measures from each context property at the same time that they arrive. Analogously, the entity states are defined by a combination of attributes and their potential values. Thus, the current state could be approached by reading the current measures for the attributes at the same time that the data is processed.

However, the time in which a certain combination of scenario and entity state occurs is not eternal, and it has a given lifespan. Within the lifespan, an indicator will use the decision criteria for a specific combination of scenario and entity state, but from the moment of some of them change (i.e. the scenario or entity state), the decision criteria need to be fitted to the new combination for the same indicator to interpret properly the new measures in its updated context.

Independently whether the data processing is carried forward using a batch or online data processing, the impact on the current scenario, and the entity state will affect the decision criteria related to a given indicator. In such a way, how to determine the current scenario and entity state is critical because it directly affects the decision-making process and their references.

4.5   MONITORING THE EVOLUTION OF ENTITIES AND SCENARIOS

The scenarios help to interpret each potential configuration of the context in which an entity under monitoring has a certain level of influence. In the same way, the states for the entity represent a way to describe each possible configuration of their attributes in a given time. As introduced in the previous section, the indicator interpretation is based on decision criteria jointly with a set of scenarios and states. This constitutes a key asset in the different big data analytical models, even considering applications focused on the energy consumption optimizing or stock markets' monitoring.

In Divan and Reynoso (2019), a proposal to approach the empirical likelihood of each scenario and entity state was introduced. Synthetically, reading measures from the data stream and matching its correspondence with an attribute or context property, it was possible to determine the current scenario and entity state related to a given entity.

Before the online monitoring on an entity starts, the scenarios and entity states had a theoretical likelihood, approached from suppositions associated with the project definition. However, once the measures started to arrive, it is possible to approach the empirical likelihoods through the occurrence frequency. That is to say, each time

TABLE 4.2

An Approach to Empirical Likelihoods Based on Frequency in the Online Monitoring

	Resting	Walking	Running	Swimming	Total
Frequency (the counter)	1,900	1,500	400	200	4,000
Relative value (an estimation)	0.475	0.375	0.100	0.050	1.000

that a state (or scenario) is identified, an accumulator is incremented in one in a vector containing the association between state (or scenario) and the counter. When it is necessary to contrast (or fit) theoretical and empirical likelihoods, a snapshot from vectors could be taken and the empirical likelihood could be obtained from the sum and division of frequencies as it is schematized in Table 4.2, using the states from Figure 4.2.

Every number in the frequency row represents the total number of processed measures per state. When it is necessary to approach the empirical likelihood, the sum of occurrences for each state is indicated under the total column. From there, a simple division between the number of measures for a given state and the number representing all the occurrences will provide an estimation (e.g. for resting state, 1,900/4,000=0.475). The same reasoning is useful for estimating the empirical likelihood of scenarios.

However, such likelihood refers to a scenario or entity state, and not their joint occurrence. To get an advantage of the data processing, a bidimensional matrix describing by the entity states on the rows, while the scenarios on the columns could be generated. Each intersection would represent the joint occurrence for a given combination between scenarios and entity states. In this way, when a conditional likelihood or joint occurrence needs to be estimated, the same reasoning described by Table 4.2 could be applied.

This is very useful in online monitoring because it provides online feedback about the possibility of occurrence of scenarios, states, and their joint occurrences. Nevertheless, it does not provide any information about the evolution of each combination of scenarios and states. That is to say, given a combination of scenarios and states, it would be very interesting to know the next combination to be reached based on its online history; in other words, to know the most probable following combination for a given scenario and entity state.

A proposal to solve this situation is to use a directed graph in which each node is associated with a new set named V obtained from the Cartesian product between the set of scenarios (i.e. SC) and entity states (i.e. ST) as it is shown in Equation 4.5.

$$\forall st \in ST \land \forall sc \in SC/V = ST \times SC \land (st, sc) \in V \tag{4.5}$$

The V set represents the directed graph vertices, while each potential transition between the vertices represents the potential transition from a given entity state and scenario to another. Be x and y a potential combination of entity states and scenarios, the E set contains all the potential transitions among (scenarios, entity states), even loops, as it is indicated in Equation 4.6.

$$E \subseteq \left\{(x,y)/(x,y) \in V^2\right\} \qquad (4.6)$$

The frequency of transitions among different combinations of scenarios and entity states could be computed using a bidimensional matrix, where the order will be determined by the size of the V set. That is to say, the matrix would need to describe a directed graph, indicating each potential origin in the rows, while each potential target is represented on the columns. Because the potential origin and targets are the same, and in other words, the origins and targets could be any element of the V set, the matrix will be $|V| \times |V|$, rewritten in terms of scenarios and entity states would be $|ST \times SC| \times |ST \times SC|$.

Articulated with the example introduced in Figure 4.2, the concept of the joint transition matrix is incorporated in Figure 4.3. As it is possible to appreciate, from the defined scenarios and entity states, it is possible to get the V set integrated by each possible combination among them; in other words, the V set is the Cartesian product from the ST and SC sets. Also, each element in V represents a node in terms of the graph theory, where it would be useful to know the probability of transiting to another node, or even, the probability to arrive to the current node from another node. Thus, a bidimensional matrix where rows and columns are integrated for each element in V is described. The bidimensional matrix allows representing in-memory a cyclic and directed graph, limiting its dimension in terms of the number of elements in the V set.

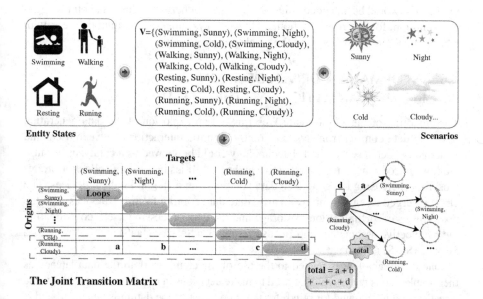

FIGURE 4.3   The role of the joint transition matrix in the online monitoring.

The principal diagonal in the matrix represents the loops for each referred node, while the rest of the intersections describe a way to transit from a given node to another. For example, the cell with an "a" letter within indicates a transition from the node "(Running, Cloudy)" to "(Swimming, Sunny)" where the indicated letter talks about the number of times in which this event has occurred.

Each node in the rows indicates the transition's origin, while each node in the columns represents the transition's target. For such a reason, the transition "(Running, Cloudy)" → "(Swimming, Sunny)" is not the same that "(Swimming, Sunny)" → "(Running, Cloudy)" due to the transition's sense is different. Both transitions could be represented in this matrix, discriminating properly each kind of transition.

In this way, the joint transition matrix is easy to compute and keep updated because each entity states and scenarios are determined on the fly, so when some transition occurs, this matrix just needs to increase the corresponding counter. It could sound naïve; however, this allows approaching the probability of transitions considering the joint occurrence among scenarios and entity states. Let supposes that the "(Running, Cloudy)" node is the current node, and you could sum all the counters in its associated row to get all the possible transitions from there. Thus, by simply dividing each counter in the mentioned row about the total sum, it would be possible to obtain each relative frequency (or empirical probability) to discover the most or fewest probable transition.

The applicability of these concepts allows deeply differentiating the big data analytics from Business Intelligence (BI) because this kind of real-time monitoring associated with the entities, scenarios, states, and associated probabilities, among other aspects, are not present in BI. The BI focuses on historical information to support the decision-making process, while in this kind of approach, big data analytics is focused on the online monitoring to anticipate situations or make decisions as soon as possible. In this sense, it imagines application environments where online monitoring would be an added value, such as green-energy production, home energy-consumption, forest fire-prevention, and waterways supervising, among others, with a high impact in the society and the whole people's health.

4.6 INFORMATION-DRIVEN DECISION-MAKING, BIG DATA ANALYTICS, AND THE CLOUD

The emerging of the World Wide Web implied an explosion of the data generation based on data coming from systems jointly with the interactions between users and systems. In such a sense, the IoT technology could be viewed as a catalyzer because it allows introducing a new perspective and possibilities in the data collection strategies, thus increasing the generated data volume from different kinds of devices. The cloud emerged as an alternative to articulate data coming from IoT devices with the possibilities of virtualization of storage and the data processing itself, consolidating the perspective associated with Infrastructure as a Service (IaaS) and Software as a Service (SaaS) (Elazhary 2019).

One of the challenges related to the use of the cloud refers to the data latency as their implications in systems that need to make decisions in real time. That is to say, the target system is outdated (or blind) for the time between the datum is generated and is consumed. When the data latency is increased (independently the reasons), the real-time systems increase the risk of making decisions based on outdated data or unknown situations (Akidau et al. 2015). This is an important aspect to be considered in different applications like stock market monitoring, fraud detection in communication networks, etc.

IoT devices are an interesting alternative to data collection, but it is worthy to mention that their local abilities (i.e. storage, processing resources, memory, etc.) are limited, and indeed, they originally were prepared to collect and transmit data. In such a context, the cloud emerged as an alternative to articulate SaaS, IaaS, and the IoT devices through different connectivity technologies (e.g. Wi-Fi, LoRa). In such a sense, a common and logical data pool was generated on the cloud from data coming on IoT devices, making them transparent for different applications. Thus, each application could apply business logic, producing and visualizing to users the information accordingly to its requirements.

Figure 4.4 describes a conceptual perspective of a cloud-based IoT architecture which could be used in a set of different applications, such as the outdoor monitoring, forest fire monitoring, and smart home, among others. In this kind of general architecture, there is a set of IoT devices distributed on the field communicated among them, continuously collecting data and sending them to the cloud through some kind of Internet connection. Data from different places are gathered on a common data pool to be processed and analyzed by an application using its business logic. In this sense, the data access is transparent for the application because it sees data as a huge data storage without any notion about how they were obtained. The application could make available its information to users through different reports or visualization strategies, but it can also notify users through some kind of alarm system about some detected situation or risk (e.g. the beginning of a potential forest fire).

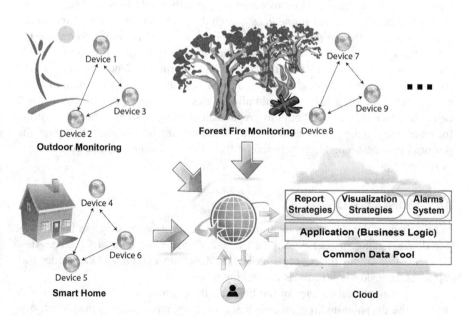

FIGURE 4.4 A conceptual perspective of a cloud-based IoT architecture.

Because of the data latency, edge computing emerged as an alternative to carry as close to users as possible the data storage and processing. The idea was to reach an articulation among the cloud and user's closest devices but incorporating some possibilities to store data and provide local answers. Thus, the device took a more active role, passing from a classical data collector to an interactive perspective. It provided a local answer by the use of the application business logic based on swarm logic. Thus, the local answer could be provided using local data, or data coming from the nearest neighbors. Communication with neighbors takes advantage of wide-range communication technologies to interact with each other. Such interaction allows collaborating through data sharing, avoiding accessing the cloud. Just in case when no enough data to answer from the local repository or nearest neighbors, the data are requested to the cloud ecosystem. This strategy tends to decrease the data latency through the increasing of local capacities which requires an increment in the hardware capacities (Khan et al. 2019).

These kinds of architectures are very promising to promote sustainable computing, distributing the processing charge, requiring lesser energy consumption, and cooling systems. This provides a context to power big data analytics applications following a swarm approach for business logic and its associated data analysis. That is to say, the devices tend to actively collaborate in the business logic contemplating its situation but also knowing the perspective of the whole system. That fosters the data interpretation from the collecting time, being able to discriminate data from information through the use of previous knowledge (e.g. using an organizational memory). Thus, this swarm schema fosters an information-driven decision-making having a special effect on measurement systems. For example, let us suppose that a certain entity is under monitoring. The measurement project is shared among each component, and the interpretation of measures through indicators jointly with scenarios, states, and their evolutions could be partially made locally. Thus, a risky situation could be detected immediately when data are read from the source, and it could be communicated to the nearest neighbors and subsequently informed to the rest of the system. As an advantage, this kind of behavior would allow it to warn all the neighbors about a given risk and collectively inform the cloud system about the risk origin and the contextual situation of their neighbors. In other words, in a forest fire, this kind of advantage would allow quickly informing the affected region, being able to notify to neighbors to prepare the defense systems, and transmitting the emergence to the cloud-based system, describing the origin, affected region, the potential propagation addresses based on the contextual situation, and current state of each neighbor.

4.7   CONCLUSION

In this chapter, the role of the measurement in the decision-making has been introduced, contrasting the perspective of the batch and online data processing. Also, differentiation among the concepts such as the data, information, and knowledge was introduced considering the decision-making process.

In this way, the emergence of the business intelligence concept in its context to support the decision-making process based on historical data was described, discriminating the perspective with the big data analytics.

Thus, the data quality affecting both measures and data was addressed to describe the different challenges and their impacts on the big data repositories. In this way, a functional proposal to discriminate different applications of big data analytics was addressed to provide a contextual reference based on its applicability.

The role of scenarios and entity states was incorporated in the measurement process jointly with its influence on the batch and real-time data processing. This allows fitting the decision criteria to evaluate a measure according to the current scenario and state, increasing the associated precision and accuracy.

A proposal to monitor the joint evolution of scenarios and entity states was proposed to approach the empirical transition likelihood. In this way, given a certain combination of (scenario, state), it would be possible to indicate the most and least likelihoods related to the following combination. This is important because it allows increasing the certainty in the multi-criteria decision-making in big data analytics.

Finally, the information-driven decision-making based on IoT devices and articulated with the cloud and fog technologies was described to provide a whole perspective about the complexity of the decision-making processing nowadays. This paradigm has taken a starring role as a real way to monitor different kinds of businesses in real time, complementing the batch data processing in terms of data analytics.

As future work, additional developments associated with the scenarios and entity states monitoring to foster better decision-making in real time will be addressed in complex contexts.

REFERENCES

Akidau, Tyler, Robert Bradshaw, Craig Chambers, Slava Chernyak, Rafael J. Fernández-Moctezuma, Reuven Lax, Sam McVeety, et al. 2015. "The Dataflow Model". *Proceedings of the VLDB Endowment* 8 (12): 1792–1803. https://doi.org/10.14778/2824032.2824076.

Anila Glory, Horison, Chandrasekaran Vigneswaran, and Shankar Sriram Shankar. 2020. "Unsupervised Bin-Wise Pre-Training: A Fusion of Information Theory and Hypergraph". *Knowledge-Based Systems*. https://doi.org/10.1016/j.knosys.2020.105650.

Artiga, Dr Marc. 2020. "Models, Information and Meaning". *Studies in History and Philosophy of Science Part C: Studies in History and Philosophy of Biological and Biomedical Sciences*, April, 101284. https://doi.org/10.1016/j.shpsc.2020.101284.

Asghari, Parvaneh, Amir Masoud Rahmani, and Hamid Haj Seyyed Javadi. 2019. "Internet of Things Applications: A Systematic Review". *Computer Networks* 148 (January): 241–261. https://doi.org/10.1016/j.comnet.2018.12.008.

Brous, Paul, Marijn Janssen, and Paulien Herder. 2020. "The Dual Effects of the Internet of Things (IoT): A Systematic Review of the Benefits and Risks of IoT Adoption by Organizations". *International Journal of Information Management* 51: 101952. https://doi.org/https://doi.org/10.1016/j.ijinfomgt.2019.05.008.

Chaudhuri, Surajit, Umeshwar Dayal, and Vivek Narasayya. 2011. "An Overview of Business Intelligence Technology". *Communications of the ACM* 54 (8): 88–98. https://doi.org/10.1145/1978542.1978562.

Chen, Yuanchang, Peter Avitabile, and Jacob Dodson. 2020. "Data Consistency Assessment Function (DCAF) ". *Mechanical Systems and Signal Processing* 141: 106688. https://doi.org/https://doi.org/10.1016/j.ymssp.2020.106688.

Chiheb, Fatma, Fatima Boumahdi, and Hafida Bouarfa. 2019. "A New Model for Integrating Big Data into Phases of Decision-Making Process." *Procedia Computer Science* 151: 636–642. https://doi.org/https://doi.org/10.1016/j.procs.2019.04.085.

Chin, Jeannette, Vic Callaghan, and Somaya Ben Allouch. 2019. "The Internet-of-Things: Reflections on the Past, Present and Future from a User-Centered and Smart Environment Perspective". *Journal of Ambient Intelligence and Smart Environments* 11 (1): 45–69. https://doi.org/10.3233/AIS-180506.

Choudhury, Suvra Jyoti, and Nikhil R. Pal. 2019. "Imputation of Missing Data with Neural Networks for Classification". *Knowledge-Based Systems* 182: 104838. https://doi.org/ https://doi.org/10.1016/j.knosys.2019.07.009.

Divan, Mario Jose, and Maria Laura Sanchez Reynoso. 2019. "Incorporating Scenarios and States Definitions on Real-Time Entity Monitoring in PAbMM". In 2019 *XLV Latin American Computing Conference (CLEI)*, 1–10. IEEE. https://doi.org/10.1109/ CLEI47609.2019.235072.

Elazhary, Hanan. 2019. "Internet of Things (IoT), Mobile Cloud, Cloudlet, Mobile IoT, IoT Cloud, Fog, Mobile Edge, and Edge Emerging Computing Paradigms: Disambiguation and Research Directions". Journal of Network and Computer Applications 128 (February): 105–140. https://doi.org/10.1016/j.jnca.2018.10.021.

Fatimah, Yun Arifatul, Kannan Govindan, Rochiyati Murniningsih, and Agus Setiawan. 2020. "A Sustainable Circular Economy Approach for Smart Waste Management System to Achieve Sustainable Development Goals: Case Study in Indonesia". *Journal of Cleaner Production*, May: 122263. https://doi.org/10.1016/j. jclepro.2020.122263.

Gao, Jianhua, Weixing Ji, Lulu Zhang, Anmin Li, Yizhuo Wang, and Zongyu Zhang. 2020. "Cube-Based Incremental Outlier Detection for Streaming Computing". *Information Sciences* 517: 361–376. https://doi.org/https://doi.org/10.1016/j.ins.2019.12.060.

Ho, Mark K., David Abel, Thomas L. Griffiths, and Michael L. Littman. 2019. "The Value of Abstraction". *Current Opinion in Behavioral Sciences* 29: 111–116. https://doi.org/ https://doi.org/10.1016/j.cobeha.2019.05.001.

Keenan, Peter Bernard, and Piotr Jankowski. 2019. "Spatial Decision Support Systems: Three Decades On". *Decision Support Systems* 116 (January): 64–76. https://doi.org/10.1016/j. dss.2018.10.010.

Khan, Wazir Zada, Ejaz Ahmed, Saqib Hakak, Ibrar Yaqoob, and Arif Ahmed. 2019. "Edge Computing: A Survey". Future Generation Computer Systems 97 (August): 219–235. https://doi.org/10.1016/j.future.2019.02.050.

Krechmer, Ken. 2016. "Relational Measurements and Uncertainty". Measurement 93 (November): 36–40. https://doi.org/10.1016/j.measurement.2016.06.058.

Krechmer, Ken. 2018. "Relative Measurement Theory: The Unification of Experimental and Theoretical Measurements". *Measurement*: Journal of the International Measurement Confederation. https://doi.org/10.1016/j.measurement.2017.10.053.

Kurnia, Parama Fadli, and Suharjito. 2018. "Business Intelligence Model to Analyze Social Media Information". *Procedia Computer Science* 135: 5–14. https://doi.org/10.1016/j. procs.2018.08.144.

Landset, Sara, Taghi M. Khoshgoftaar, Aaron N. Richter, and Tawfiq Hasanin. 2015. "A Survey of Open Source Tools for Machine Learning with Big Data in the Hadoop Ecosystem". *Journal of Big Data*. https://doi.org/10.1186/s40537-015-0032-1.

Laukkanen, Minttu, and Nina Tura. 2020. "The Potential of Sharing Economy Business Models for Sustainable Value Creation". *Journal of Cleaner Production* 253: 120004. https://doi.org/https://doi.org/10.1016/j.jclepro.2020.120004.

Lee, In. 2019. "The Internet of Things for Enterprises: An Ecosystem, Architecture, and IoT Service Business Model". *Internet of Things* 7: 100078. https://doi.org/https://doi. org/10.1016/j.iot.2019.100078.

Lemoine, Frédéric, Tatiana Aubonnet, and Noëmie Simoni. 2020. "Self-Assemble-Featured Internet of Things". *Future Generation Computer Systems* 112: 41–57. https://doi.org/ https://doi.org/10.1016/j.future.2020.05.012.

Li, Pengfei, Kezhi Mao, Yuecong Xu, Qi Li, and Jiaheng Zhang. 2020. "Bag-of-Concepts Representation for Document Classification Based on Automatic Knowledge Acquisition from Probabilistic Knowledge Base". *Knowledge-Based Systems* 193 (April): 105436. https://doi.org/10.1016/j.knosys.2019.105436.

Luo, Xihaier, and Ahsan Kareem. 2020. "Bayesian Deep Learning with Hierarchical Prior: Predictions from Limited and Noisy Data". *Structural Safety* 84: 101918. https://doi.org/https://doi.org/10.1016/j.strusafe.2019.101918.

Martins, Carmen, Ana Salazar, and Alessandro Inversini. 2015. "The Internet Impact on Travel Purchases: Insights from Portugal". *Tourism Analysis* 20 (2): 251–258. https://doi.org/10.3727/108354215X14265319207632.

Merino, Jorge, Ismael Caballero, Bibiano Rivas, Manuel Serrano, and Mario Piattini. 2016. "A Data Quality in Use Model for Big Data". Future Generation Computer Systems 63: 123–130. https://doi.org/https://doi.org/10.1016/j.future.2015.11.024.

Plattner, Hasso. 2009. "A Common Database Approach for OLTP and OLAP Using an In-Memory Column Database". In *Proceedings of the* 35th *SIGMOD International Conference on Management of Data - SIGMOD '09*, 1–2. New York, NY: ACM Press. https://doi.org/10.1145/1559845.1559846.

Ramalho, Felipe Diniz, Petr Ya. Ekel, Witold Pedrycz, Joel G. Pereira Jr., and Gustavo Luís Soares. 2019. "Multicriteria Decision Making under Conditions of Uncertainty in Application to Multiobjective Allocation of Resources". Information Fusion 49: 249–261. https://doi.org/https://doi.org/10.1016/j.inffus.2018.12.010.

Ramli, Muhammad Rusyadi, Philip Tobianto Daely, Dong-Seong Kim, and Jae Min Lee. 2020. "IoT-Based Adaptive Network Mechanism for Reliable Smart Farm System". *Computers and Electronics in Agriculture* 170: 105287. https://doi.org/https://doi.org/10.1016/j.compag.2020.105287.

Sachdev, Sumeet, Joel Macwan, Chintan Patel, and Nishant Doshi. 2019. "Voice-Controlled Autonomous Vehicle Using IoT". Procedia Computer Science 160: 712–717. https://doi.org/https://doi.org/10.1016/j.procs.2019.11.022.

Safhi, Hicham Moad, Bouchra Frikh, and Brahim Ouhbi. 2019. "Assessing Reliability of Big Data Knowledge Discovery Process". Procedia Computer Science 148: 30–36. https://doi.org/10.1016/j.procs.2019.01.005.

Sahal, Radhya, John G. Breslin, and Muhammad Intizar Ali. 2020. "Big Data and Stream Processing Platforms for Industry 4.0 Requirements Mapping for a Predictive Maintenance Use Case". *Journal of Manufacturing Systems* 54 (January): 138–151. https://doi.org/10.1016/j.jmsy.2019.11.004.

Song, Jaeki, and Fatemeh Mariam Zahedi. 2006. "Internet Market Strategies: Antecedents and Implications". *Information & Management* 43 (2): 222–238. https://doi.org/10.1016/j.im.2005.06.004.

Sterling, Ryan, and Cynthia LeRouge. 2019. "On-Demand Telemedicine as a Disruptive Health Technology: Qualitative Study Exploring Emerging Business Models and Strategies Among Early Adopter Organizations in the United States". *Journal of Medical Internet Research* 21 (11): e14304. https://doi.org/10.2196/14304.

Yaqoob, Ibrar, Ibrahim Abaker Targio Hashem, Abdullah Gani, Salimah Mokhtar, Ejaz Ahmed, Nor Badrul Anuar, and Athanasios V. Vasilakos. 2016. "Big Data: From Beginning to Future". *International Journal of Information Management* 36 (6): 1231–1247. https://doi.org/10.1016/j.ijinfomgt.2016.07.009.

Zhao, Jun, and John Serieux. 2020. "Economic Globalization and Regional Income Convergence: Evidence from Latin America and the Caribbean". *World Development Perspectives* 17: 100176. https://doi.org/10.1016/j.wdp.2020.100176.

5 Performance Analysis for Provisioning and Energy Efficiency Distributed in Cloud Computing

R. Ganesh Babu
SRM TRP Engineering College

D. Antony Joseph Rajan
SCSVMV University

Sudhanshu Maurya
Graphic Era Hill University

P. Jayachandran
Thiruthangal Nadar College

CONTENTS

5.1 INTRODUCTION: BACKGROUND AND DRIVING FORCES

The huge point of convergence of the present world is to reduce the nuances of the World Wide Web market update estimates. With its organizations on establishment, stage, and programming, dispersed figuring is the best guide to the IT market [1]. Disseminated registering is a way for growing the capacities of an item/gear without applying extra hypotheses on it [2]. It contributes to the expansion of the information and resources that are already available on it. IBM's elegant business enterprises for IT upgrades. Disseminated registration is a method of increasing an item's/

DOI: 10.1201/9781003032328-5

capabilities gear's without putting any more money into it. It contributes to expanding the information and resources available on it from now on [3].

At this time, the open doors for cloud providers have been extended for the promotion [4]. There are various issues confronted by a cloud supplier when offering services to the consumers of an organization, among which security perspectives, energy efficiency, and cost are noteworthy ones. The cloud gives many advantages such as snappy plan, lower costs, flexibility, quick provisioning, brisk adaptability, and inescapable frameworks to achieve a workable pace, quality, hypervisor protection against composed ambushes, recovery of negligible exertion failure, data amassing game plans, organized security, and ceaseless and easy identification of dangers that are possible to occur in an organization [5]. Security in the cloud server ranch may face many noteworthy challenges, and hence the securing the enormous frameworks from a variety of attacks is essential.

A great deal of research is leaving to the nearby silent altogether increasingly secure framework required in the current situation. Various are verified under security and provide a variety of them: Intensity of security organization, security of the server ranch plane, security since the DDOS ambushes and the security ISP's, 76% of IT managers and the CIO's alluded to security as summit test deflecting their gathering the cloud organizations sculpt name Clavister [6]. With the expansion of the calculating resources, there is a significant need for assistance in managing the imperativeness consumed by servers. The proposal of the green figuring has created another ground in the zone of imperativeness protection [7].

Green green Computing insinuates the imperativeness gainful piece of appropriated registering that believe the carbon radiation thought, natural – essentialness considerations in IT types of apparatus, and green middle people. The cost spent for achieving imperativeness is the major part of expenditure in preparing a cloud platform. With the expansion in the solicitations of imperativeness ICT (Information correspondences development) part, the charge of essentialness is additionally growing while achieving the decrease in the ordinary possessions. The assumption for net essentiality use, which includes net total imperativeness use from all devices as soon as possible, is essentially high. Moreover, the green handling supports various features such as virtualization, server utilization, multi-inhabitance, dynamic provisioning, and data center efficiency [8].

5.2   EFFICIENCY OF ENERGY CONSERVATION IN CLOUD COMPUTING

The Energy–Efficient arrangements perform indistinguishable tasks as before while consuming less energy. Vitality protection is retreating or abandoning a support of spare vitality. It is numerous associated with vitality operation at various levels such as equipment, servers, other system gadgets, various wired and remote systems, and so on [9]. GHGs and CO_2 reduce essentialness use at server ranch level. In relation to the issue of force reduction, the use of maker control in VM's can be restricted at the organization stage. Similarly, the maker shows it probably with a stack-changing DVFS and booking technique. The maker wraps up with an exhibited proposed technique that diminishes the force use and increases advantages [10].

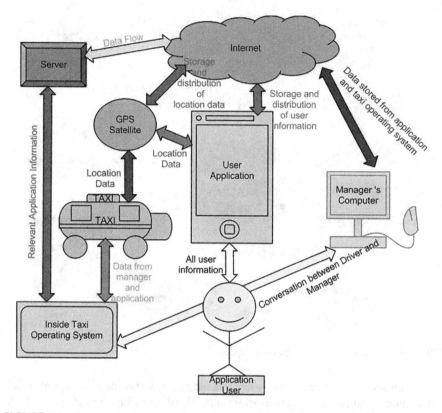

FIGURE 5.1 Energy efficiency conversation in cloud computing.

Figure 5.1 depicts a common estimation for the development of weight to a virtual machine. If the maker discovers issues such as underutilization and overconsumption of benefits, then they improve advantages by reducing the cost for moving data. The implemented EEA (Energy-Efficient Algorithm) and its differences from currently available computations. In order to discover how much energy is consumed by a typical server farm, exact now investigation r focuses around exceptional weight measurement, virtualization, and customized control the board. The maker discusses the server and the switch procedure equally for imperativeness adequacy. Message line at the switches and remote transmission arranging frameworks are projected as a response for various circumstances. In order to clarify the energy cells, the manufacturer has made available a number of server ranches that are limited by vitality units rather than control. Learn about various problems with energy units, the circumstances of intensity devices at various levels, and the cost of firewood rack [11].

5.3 DATA SECURITY IN CLOUD NETWORK

The term "security in flowed enrolling" refers to information security in cloud sorting. As more devices are connected to the cloud, configure information consistency and security to be difficult and central [12]. To identify any assault while adjusting

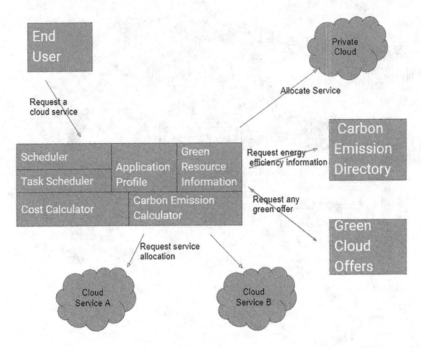

FIGURE 5.2   Data security in cloud computing.

movements of data, the entire driving force should be checked at accomplishment of each contraption and different protective measures should be in use for information security in the system. It clarified the movement and position of security components and arranged basic issues faced while giving security in the cloud condition [13]. The considered various security issues at various levels of assistance, particularly in the same way that they discussed security issues with the authorities [14]. She considers with provides beneficial security approaches. The model appears to be focused on providing information security handling and determining a good rate for meta-information data [15].

Figure 5.2 shows careful provisioning of protected imperativeness to dispersed registering resources in virtualized stages. It mainly discusses the shirking from the server ranch level [16]. The energy-efficient cloud security in a circumstance is exceptionally inconvenient, and it fundamentally requires the following expectations. Furthermore, significant parameters such as a decrease in CO_2 outpouring security in data middle level as different methodology can be projected to offer a solution in both problems [17].

5.4   COMPARATIVE STUDY OF EXISTING TECHNIQUES

A systems segment typically uses cloud computing, load balancing, virtualization, and preparation [18]. These methodologies are satisfactorily valuable in different activities connected to disseminate registering. With brought together circulated processing, virtualization is a guideline.

5.4.1 PROBLEM IDENTIFICATION

The essentialness usage of the systems is connected with various cash-related, regular and structure execution. Earth Simulator, for example, with two HPCS Petaflop systems that use 12 and 10 0 megawatts apex control exclusively. With an unfavorable cost of $100 per megawatt, essential costs throughout peak action times for 1,200, and 11,000 dollars used consistently and independently; the acceptable spending diagram of the many HPCS overseers. The increasing temperature of the circuit not simply crashes the circuit, but it also reduces the lifetime of its sections [19].

5.5 IMPLEMENTATION

The issue of centrality use in pervasive getting ready (HPC) structures has pulled in a huge load of thought. Taking this into consideration, various importance cautious assessments made in unique layers of framework productive running the construction may in a circuitous way save more noteworthy importance [20].

At present time, assessments are made in two layers. Irrespective, creativity and critical thinking three evaluations with really. Second, it suggests two evaluations for the appearance and asset provisioning of Map decrease applications (an uncommon parametric scattered construction beginning at now utilized by Google, Yahoo, Face book and LinkedIn) in context on its game-plan boundaries. Certainly, evaluating the presentation (e.g., execution time or CPU clock ticks) of a Map Reduce application can be subsequently utilized for unbelievable orchestrating of such applications in hazes or get-togethers [21].

Through lone recollections, an equivalent figuring framework is incorporated into N homogeneous of the processors. In the structures ,exchanging time among frequencies overlooked securely in the processors considering the way of time being used to modify starting with recurrent then the accompanying (40–160 μs) is totally extra minor than the execution season of tasks (in any event 1 ms).

A huge load tasks tended to $A^{(1)}, A^{(2)} ..., A^{(M)}$ is by a coordinated non-cyclic endeavor chart is in addition thought to execute HPC framework. Here, the kth-task $\left(A^{(k)}\right)$ consolidates with the five boundaries: $T^{(k)}$ is the entire processor time assigned for undertaking of execution with slack time; $t_i^{(k)}$ is the execution of time for utilizing rehash; The critical number of clock f_j demonstrate the project $f_{ideal}^{(k)}$ necessities in the ideal vigorous recurrent dependent, $K^{(k)}$ which comprehends the ideal centrality utilization for execution of the time processor exhaust on the implement $t_{os}^{(k0)}$ the task in the chief arranging

DVFS-empowered processors execute an endeavor using a separate course of action of energy repeat sets $\left(f_i, v_i\right)$ in which $\left(v_1, v_2 ..., v_N\right)$ and $\left(f_1, f_2 ..., f_N\right)$. In CMOS-based processors, the force use of a processor contains two areas: (i) an extraordinary part that is essentially related to CMOS circuit trading imperativeness,and (ii) a static part that keeps an eye on the circuit CMOS control. The whole authority consumption $\left(P_d\right)$ is unsurprising as follows:

$$\begin{cases} P_d = \lambda f v^2 + \mu v \\ f\alpha \dfrac{(v-v_t)^2}{v_t} \end{cases} \tag{5.1}$$

$$\text{If } (f_i, v_i) < (f_i, v_i) \rightarrow P_d(f_i, v_i) < P_d(f_i, v_i) \tag{5.2}$$

$$E^{(k)} = P_d t_i^{(k)} + P_l \left(T^{(k)} - t_i^{(k)} \right) \tag{5.3}$$

$$P_d = \lambda f^3 + \gamma \tag{5.4}$$

$$E^{(k)}(f_1, t_1) < E^{(k)}(f_2, t_2) \tag{5.5}$$

RDVFS Algorithm: Slack Reclamation by One Frequency
Input: The scheduled tasks on a set of P Processor

1. For task $A^{(k)}$ scheduled on processor P_j
2. Compute the optimum continuous frequency $f_{\text{Opt}}^{(k)} - \text{cont.}$ from Equation 5.5
3. Pick the closest higher frequency to $f_{\text{Opt}}^{(k)} - \text{cont.}$ in the CPU frequency set

$$\text{e.g.,} \quad \left. \begin{array}{c} [f_{\max} > \cdots > f_n > f_{n-1} f_{\min}] \\ f_n > f_{\text{Opt}}^{(k)} - \text{cont.} > f_{n-1} \end{array} \right\} \Rightarrow f_{\text{RDVFS}}^{(k)} = f_{n-1}$$

4. $t_{\text{RDVFS}}^{(k)} = \dfrac{f_{\text{Opt}}^{(k)} - \text{cont.}}{f_{\text{RDVFS}}^{(k)}} T^{(k)}$

5. $E_{\text{RDVFS}} = f_{\text{RDVFS}}^{(k)} t_{\text{RDVFS}}^{(k)} + P_{\text{idle}} \left(T^{(k)} - t_{\text{RDVFS}}^{(k)} \right)$

6. end for
7. Return $\left(f_{\text{RDVFS}}^{(k)}, t_{\text{RDVFS}}^{(k)} \right)$ for all task

RDVFS has been redesigned in terms of the count displayed by control adaptable predominant gatherings supporting for the DVFS. It diminishes the essentialness use of the processors with picking the humblest open processor $\left(f_{\text{RDVFS}} \right)$ recurrence fit for completing an undertaking in a given time period. The subtleties of RDVFS calculation are appeared for each errand allotted to a processor, $f_{\text{RDVFS}}^{(k)}$ which is the primary recurrence bigger than ideal recurrence $\left(f_{\text{Opt}}^{(k)} - \text{cont.} \right)$ determined from Equation 5.3, is probably going to be the greatest discrete frequent candidate to the execution endeavor within the known timeframe, and the covering is to be connected slack time. As referenced beforehand, a noteworthy containment of RDVFS strategy is utilization of only solitary repeat to implement the endeavor [22].

In this space execution of MMF-DVFS, booking stood apart from different ways of thinking in all cases. Figures 5.3 and 5.4 show that paying little mind to the way that the ampleness considering everything,checking MMF-DVFS,in decrease centrality in LU decay is basic,those calculations is less fortunate presentation on

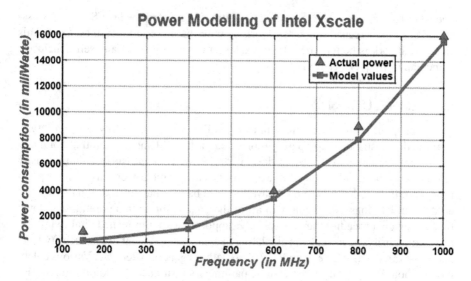

FIGURE 5.3 The least-square modeling of Intel Xscale.

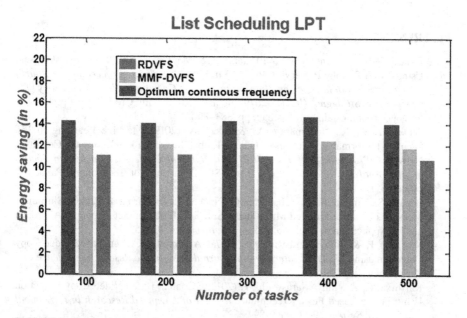

FIGURE 5.4 Number of tasks vs energy saving (in %).

Gauss-Jordan tasks in Figures 5.3 and 5.4 of checks can effectively lessen centrality utilization [23].

Expanding the measure of processors helps inarranging time and thusdecreases the length of time; notwithstanding,as weight,it additionally gathers the framework time. Kind of four monitors the level of everything is done vitality saving of design taking into account the number of processors for self-decisive and LU separating undertaking diagrams. The important limitations of most of the DVFS-based assessments that occur recurrently while working with the mixes are yet to be fixed. This calculationwork is better when the workstation runsvery well at any self-conclusive gameplan of frequencies [24].

5.6 CONCLUSION

In this chapter, the requirementof intensity utilization with vitality productivity is examined in distributed computing reproduction. It has demonstrated that not many significant segments of the cloud design can be used for measuring high-intensity scattering in the cloud. The possible approaches to assemble every division for planning a vitality productivity model have likewise been contemplated. Finally discovered the utilizations evaluation with all of the enquiry up to this point, break down that no current procedures are completely supported with diminished vitality utilization and security. As a result, a calculating decreasing increment use to security level keeps the information safe from RDVFS assaults is generally terrible. Various calculations should be prepared that include parameters insurance in addition to RDVFS as saults; decrease of RDVFS information focal level; planning on virtual machine allowing for the vitality factor.

REFERENCES

1. Ganesh Babu, R., Antony Joseph Rajan, D., & Maurya, S. (2019). Multi Data Users Using Novel Privacy Preserving Protocol in Cloud Networks. *Proceedings of Fourth International Conference on Information and Communication Technology for Competitive Strategies (ICTCS-2019)* in association with CRC Press (Taylor and Francis Group), Udaipur, Rajasthan, pp. 883–890.
2. Ganesh Babu, R., Nedumaran, A., & Sisay, A. (2019). Machine Learning in IoT Security Performance Analysis of Outage Probability of Link Selection for Cognitive Networks. *Proceedings of Third IEEE International Conference on IOT in Social, Mobile, Analytics and Cloud (I-SMAC)*, SCAD Institute of Technology, Palladam, India, pp. 15–19.
3. Begam, S., Vimala, J., Selvachandran, G., Ngan, T.T., & Sharma, R. (2020). Similarity Measure of Lattice Ordered Multi-Fuzzy Soft Sets Based on Set Theoretic Approach and Its Application in Decision Making. *Mathematics, 8*, 1255.
4. Karthika, P., & Vidhya Saraswathi, P. (2017a). A Survey of Content based Video Copy Detection using Big Data. *International Journal of Scientific Research in Science and Technology, 3*(5), 114–118.
5. Karthika, P., & Vidhya Saraswathi, P. (2017b). Content Based Video Copy Detection Using Frame Based Fusion Technique. *Journal of Advanced Research in Dynamical and Control Systems, 9*(17), 885–894.
6. Karthika, P., & Vidhya Saraswathi, P. (2018). Digital Video Copy Detection Using Steganography Frame Based Fusion Techniques. *Proceedings of the International Conference on ISMAC in Computational Vision and Bio-Engineering in Association with Lecture Notes in Computational Vision and Biomechanics*, Vol. 30, Springer, Cham, pp. 61–68.
7. Vo, T., Sharma, R., Kumar, R., Son, L.H., Pham, B.T., Tien, B.D., Priyadarshini, I., Sarkar, M., & Le, T. (2020). Crime Rate Detection Using Social Media of Different Crime Locations and Twitter Part-of-speech Tagger with Brown Clustering, 4287–4299.

8. Nguyen, P.T., Ha, D.H., Avand, M., Jaafari, A., Nguyen, H.D., Al-Ansari, N., Van Phong, T., Sharma, R., Kumar, R., Le, H.V., Ho, L.S., Prakash, I., & Pham, B.T. (2020). Soft Computing Ensemble Models Based on Logistic Regression for Groundwater Potential Mapping. *Applied Science, 10*, 2469.

9. Karthika, P., Ganesh Babu, R., & Nedumaran, A. (2019). Machine Learning Security Allocation in IoT. *Proceedings of IEEE International Conference on Intelligent Computing and Control Systems (ICICCS)*, Vaigai College of Engineering, Madurai, India, pp. 474–478.

10. Jha, S. et al. (2019). Deep Learning Approach for Software Maintainability Metrics Prediction. *IEEE Access, 7*, 61840–61855.

11. Sharma, R., Kumar, R., Sharma, D.K., Son, L.H., Priyadarshini, I., Pham, B.T., Bui, D.T., & Rai, S. (2019). Inferring Air Pollution from Air Quality Index by Different Geographical Areas: Case Study in India. *Air Quality, Atmosphere, and Health, 12*, 1347–1357.

12. Mekonnen, W.G., Hailu, T.A., Tamene, M., & Karthika, P. (2020). A Dynamic Efficient Protocol Secure for Privacy Preserving Communication Based VANET. *Proceedings of Second International Conference on Computational Intelligence in Pattern Recognition (CIPR) in Association with Advances in Intelligent Systems and Computing*, Vol. 1120, Springer, Singapore, pp. 383–393.

13. Nedumaran, A., & Jeyalakshmi, V. (2014). PARURP: Power Aware and Route Utility Function Based Routing Algorithm for Mobile Ad Hoc Networks. *International Journal of Innovative Science Engineering and Technology, 1*(9), 427–430.

14. Nedumaran, A., Ganesh Babu, R., Kass, M.M. & Karthika, P. (2019). Machine Level Classification Using Support Vector Machine. *AIP Conference Proceedings of International Conference on Sustainable Manufacturing, Materials and Technologies (ICSMMT)*, Coimbatore, India, Vol. 2207, Issue. 1, pp. 020013–1–020013–10.

15. Sharma, R., Kumar, R.,Singh, P.K.,Raboaca, M.S.,&Felseghi, R.-A.(2020). A Systematic Study on the Analysis of the Emission of CO, CO_2 and HC for Four-Wheelers and Its Impact on the Sustainable Ecosystem. *Sustainability, 12*, 6707.

16. Nedumaran, A., Jeyalakshmi, V., &Girmu birru, D. (2019). Link Stability for Energy Aware Efficient Multicast Routing Algorithm Using MANET. *International Journal of Recent Technology and Engineering, 8*(1S2), 293–297.

17. Sharma, S.et al.(2020). Global Forecasting Confirmed and Fatal Cases of COVID-19 Outbreak Using Autoregressive Integrated Moving Average Model.*Frontiers in Public Health*. https://doi.org/10.3389/fpubh.2020.580327.

18. Hailu, T.A., & Nedumaran, A. (2019). A Survey on Provisioning of Quality of Service (QoS) in MANET. *International Journal of Research and Advanced Development, 3*(2), 34–40.

19. Malik, P. et al. (2021). Industrial Internet of Things and Its Applications in Industry 4.0: State-of the Art.*Computer Communication, 166*, 125–139, Elsevier.

20. (2020). Analysis of Water Pollution Using Different Physico-Chemical Parameters: A Study of Yamuna River. *Frontiers in Environmental Science*. https://doi.org/10.3389/fenvs.2020.581591.

21. Dansana, D. et al. (2021). Using Susceptible-Exposed-Infectious-Recovered Model to Forecast Coronavirus Outbreak. *Computers, Materials & Continua, 67*(2), 1595–1612.

22. Vo, M.T., Vo, A.H., Nguyen, T., Sharma, R., & Le, T. (2021). Dealing with the Class Imbalance Problem in the Detection of Fake Job Descriptions. *Computers, Materials & Continua, 68*(1), 521–535.

23. Sachan, S., Sharma, R., & Sehgal, A. (2021). Energy Efficient Scheme for Better Connectivity in Sustainable Mobile Wireless Sensor Networks. *Sustainable Computing: Informatics and Systems, 30*,100504.

24. Ghanem, S., et al. (2021). Lane Detection under Artificial Colored Light in Tunnels and on Highways: An IoT-Based Framework for Smart City Infrastructure. *Complex & Intelligent Systems*. https://doi.org/10.1007/s40747-021-00381-2.

6 Using Internet of Things (IoT) for Smart Home Automation and Metering System

R. Ganesh Babu
SRM TRP Engineering College

J. Bino
St. Joseph's Institute of Technology

S. Kalimuthu Kumar
Kalasalingam Academy of Research and Education

K. Elangovan
Siddharth Institute of Engineering and Technology

G. Manikandan
Dr. M.G.R Educational and Research Institute

CONTENTS

6.1 INTRODUCTION

Improving creativity makes everyday human life simpler, with the aid of shrewd structures recently created. Due to the rapid development in technology creativity and cleverly implemented structures, individuals are increasingly fascinated to include various types of gadgets in the use of the internet. Network of stuff makes the innovation world an innovative breakthrough with another period of knowledge

DOI: 10.1201/9781003032328-6

development [1,2]. IoT can be described as a mixture of various kinds of gadgets such as advanced cells, resulting in a new kind of interaction between objects and individuals and with the objects [3]. IoT's main objective is to monitor all sorts of electrical items or devices around us in an even simpler, suggestive and smoother manner [4]. IoT aims to boost the perceptible efficiency of electrical equipment by claiming smart living with safety, health and diversion. IoT engineering is used as a starting point for shrewd homes and amazing growth to raise the quality of daily comforts of life [5].

The idea of a shrewd home is an increasing excitement amongst buyers as of late. Work bunches on home robotization, like IoT, are under way. It have set up a good home environment with high protection and low cost of use using IoT [6]. In the Indian environment, it proposed an IoT-based home computerization using minimal effort on Android phones [7]. It has developed a system that uses IoT for power use and security in households of great interest. In this case, it used a framework for managing images to interpret human exercises. With the assistance of Wi-Fi and GSM innovation [8], it designed a home robotization system that could monitor diverse family gadgets. While different studies are taking place, there is still the degree to develop this home mechanization process using IoT [9].

6.2   SYSTEM ARCHITECTURE

The structure square map appears in Figure 6.1. A platform was built in that context. The client first gets from where the person in question is going to send an order over the Internet. At that point, the microcontroller will get the order through a Wi-Fi unit. From that point on, the electronic gadgets can be function as per the order sent by the customer. The framework includes correspondence coming from both directions [10]. The first is to track the gadgets that use the web, as recently delineated. The last one is that the customer will get details from the web about the use of force

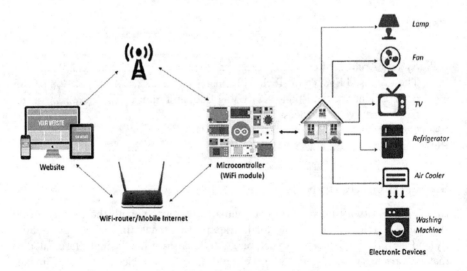

FIGURE 6.1   System architecture.

by the gadgets as the perusing meter can be legitimately obtained from the web [11]. That finances the gadgets transmitting details about the use of force to the site using a microcontroller over Wi-Fi. The customer can see the meter perusing and the data billing from the site [12].

From the outset, the client will choose the brilliant alternative house. At that point, the consumer will be redirected to the page from where it is possible to monitor home appliances such as fan, lamp, TV, air cooler, laundry washer, and so on. Shrewd home equipment photo appears [13]. The date, time, and temperature are constant information which means that each of these information is collected directly from the web [14,15]. Another piece of this system appears, which is a meter perusing frame option which implies the use of force of the gadgets and the charging data. A meter number will be issued for an issued client. Where the consumer can get to the data from the gadgets about the genuine force, power factor, obvious force, voltage supply, and vitality utilization. In addition, the client can also obtain the charging details from the web. The website is stepped up at regular intervals with the intention that customers can get the latest update of their meter. During this vigorous time, only 3 kb/s of information was used. Therefore the consumer does not need to use anIoT of site details [16].

6.3 THE PROPOSED ENCRYPTON TECHNAL PROPERTY

For preparing a calculation, three significant elements must be considered:

- The calculation must be sufficiently simple to be easily and accurately measured and examined.
- An encrypter shall provide more security edge than the requisite incentive against the known assaults.
- Well-known, widely tested and accurate methods and ideas for the framework must be used.

As demonstrated by the previously mentioned centers, utilizing the blend of altered Advanced Encryption Standard (AES) figuring and Arnold pandemonium mapping, an image encryption computation is proposed here which is proficient from both security and speed viewpoint. This paper exploits the general structure of standard AES computation. A couple of alterations were made to make the proposed procedure worthy for the encryption of the documents [17]. These terms incorporate two upgrades to the first AES encryption count: the main change is to supplant the proposed spread exercises with permutable activity in the ordinary encryption computation, and the resulting change is to supplant the straight change with the bit fuse movement. This represents the proposed encryption strategy square outline. The nuances of the proposed count are clarified in the accompanying area a little bit at a time [18,19].

6.3.1 GENERATE THE ROUND KEY USING BEDLAM FRAME

In the projected encryption calculation, Arnold bedlam frame is used to generate the key. Expect the first picture to be of M/N size and require n adjustments for encryption. In this way, $n+2$ assortments of N/M scope are produced utilizing Equations 6.1 and 6.2. Every cluster represents the round key of the CCAES calculation where the customer can get the data from the gadgets about genuine force, power factor, obvious force, supply voltage, and vitality utilization. Therefore, the customer can get the billing details from the web. The platform is invigorated at regular intervals with the intention that client will be able to peruse the latest update of its meter. During this vigorous time, only 3 kb/s of information was used. Therefore the consumer does not need to use a lot of site details.

$$X1_{n+1} = \text{mod}\big((X1_n + (a \times X2_n), 256)\big) \tag{6.1}$$

$$X2_{n+1} = \text{mod}\big(b \times X1_n + ((a \times b + 1) \times X2_n), 256\big) \tag{6.2}$$

$$K(j,k,i) = \text{floor}\Big(\text{mod}\big((k(j,k,i) * (10^{14})), 256\big)\Big) \tag{6.3}$$

Here, $X1 = 0.0215$, $X2 = 0.5734$, $a = 255.9998$, and b is equivalent to0.

6.3.2 ENCRYPTION PROCESS

Stage 1: The starting image is positioned in a state lattice with a similar dimension. On this grid all operations are carried out.
Stage 2: Second, the state grid bytes are XOR with zero-round key compared bytes, and the set1 is round value.
Stage 3: To round =1: round number.

The summation of pixels the state grid is obtained above all the underlying estimate of sum =0.

$$\text{Sum} = \sum_{i=1}^{N} \sum_{j=1}^{M} \text{State}(i,j) \tag{6.4}$$

Equation 6.4, if rounds = even; $i = 1, j = 1$

$$\text{Sum} = \text{Sum} - \text{State}(i,j) \tag{6.5}$$

$$V = \text{floor}\left(\text{mod}\left(\frac{\text{Sum}}{256^5} \times 10^{10,256}\right)\right) \tag{6.6}$$

The structure1010 is utilized, and afterward, the rest of the division to256 is determined where the guess of the number created is inside the pixel estimation extent of the picture.
In case $A_0 = 124$, $i = 1$, and $j = 1$:

$$\text{State}(i,j) = \text{State}(i,j) \oplus v \oplus A_0 \tag{6.7}$$

$$\text{State}(i,j) = \text{State}(i,j) \oplus v \tag{6.8}$$

Conditions (4) and (5) are rehashed on the off chance that changes = odd;

Stage 4: At that point, $I=N$ and $j=M$ on the off chance: $A_0 = 124$, and Equation6.6; in any case, an Equation6.7 a rehash.

Stage 5: It cyclic movement of columns: every column of the state lattice is given cyclic movement of one side, not exactly the column count.

Stage 6: Linear transformation activity: The picture < denotes the cyclic movement, and the popular movement and XOR operation are represented separately by both. This activity is carried out with state grid information, measuring 16 bytes. It implies that the immediate activity is completed on the underlying 16 bytes and in this way on the second 16 bytes and is kept up for the whole state matrix.

Stage 7: Round key XORed the state grid, and round value extends one row. Heading to stage 3 on the off chance that round <= n.

Stage 8: The last round key is XORed on the State grid. Taking into account the steps referenced, the present state structure is the blurred representation of the proposed technique (Figure 6.2).

The method of decoding is similar to that used for encryption. The key difference is that it conversely performs a few steps. At first, the unscrambling method renders key it mapping Arnold, and it amount of focus is set at 10, and afterward, the deciphering activity starts. The inversion of the XOR activity is equivalent to the XOR operation is equal to the XOR operation, and it is enough to eventually dispose [3]. Here, XOR changes to XOR the encoded image with the last word. The reverse of the proposed encryption

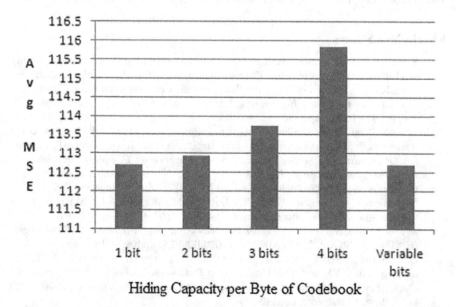

FIGURE 6.2 Hiding capacity versus average Maxillary Skeletal Expansion (MSE).

calculation is performed at that point; this operation encoded at the key first image is XORed. The converse of the direct shift operation is added to the figure picture in the corresponding advance [20]. In the subsequent step, the task of moving the column backwards is performed that a column in the activity of encryption is swung to one side by n in action, the translation of a similar line *n* units shifted with one side. A condition is verified in the last advance; this condition is with the end goal that the quantity of the rounds is odd the last pixel and something else, performed the spreading activity [21,22].

In terms of the quantity of rounds this process is continued. The picture from past developments at that stage is XORed by the primary key. The picture that has come about is equal to the first one. The beneficiary simply requires the underlying qualities and can use these qualities to get to the comparative key of the encryption system. During the current experiments, various disarray structures were investigated. The Arnold turmoil framework's effects have been superior to other systems. That is why this $\text{State}(i,j) = \text{State}(i,j) \oplus v$ uses Arnold's disarray system to construct the hidden key era pseudorandom arrangements [23].

6.4 CONCLUSION

A shrewd home computerization platform is implemented in this chapteralongside metering platform that uses web of things. The chapter's fundamental goal is to monitor the home machines via the web just like electronic gadgets. The site's UI is designed in an incredibly simple manner and can be used by all. Not only can the customer monitor and track the electronic devices with the aid of this platform, but they can also watch the metering system. Furthermore, the power provider will screen the metering system to see if there is any ambiguity in the dispersion picture. The platform can also be used to monitor the power tabs on the internet.

REFERENCES

1. Karthika, P., Ganesh Babu, R., Karthik, P.A.: Fog Computing Using Interoperatibility and IoT Security Issues in Health Care. In: SharmaD.K., BalasV.E., SonL.H., SharmaR., CengiK. (eds).*Proceedings of Third International Conference on Micro-Electronics and Telecommunication Engineering.* Lecture Notes in Networks and Systems, vol. 106, pp. 97–105.Springer, Singapore (2019).
2. Ganesh Babu, R., Karthika, P., Aravinda Rajan, V.: Secure IoT Systems Using Raspberry Pi Machine Learning Artificial Intelligence. In: SmysS., SenjyuT., LafataP. (eds).*Proceedings of Second International Conference on Computer Networks and Inventive Communication Technologies.* Lecture Notes on Data Engineering and Communications Technologies, vol. 44, pp. 797–805.Springer, Singapore, 2020.
3. Nedumaran, A., Ganesh Babu, R., Kass, M.M., Karthika, P.: Machine Level Classification Using Support Vector Machine. *AIP Conference Proceedings of International Conference on Sustainable Manufacturing, Materials and Technologies (ICSMMT 2019)*, vol. 2207, pp.020013-1–020013-10, Coimbatore, India (2020).
4. Ganesh Babu, R., Karthika, P., Elangovan, K.: Performance Analysis for Image Security using SVM and ANN Classification Techniques. In: *Third IEEE International Conference on Electronics, Communication and Aerospace Technology*, pp. 460–465. IEEE Press, Coimbatore, India (2019).

5. Karthika, P., Vidhya Saraswathi, P.: Image Security Performance Analysis for SVM and ANN Classification Techniques. *International Journal of Recent Technology and Engineering*8(4S2), 436–442 (2019).

6. Ganesh Babu, R., Nedumaran, A., Sisay, A.: Machine Learning in IoT Security Performance Analysis of Outage Probability of Link Selection for Cognitive Networks. In: *Third IEEE International Conference on IOT in Social, Mobile, Analytics and Cloud (I-SMAC)*, pp. 15–19. IEEE Press, Palladam, India (2019).

7. Karthika, P., Hailuand, T.A., Mekonnen, W.G.: Enhancing the Performance Analysis of Interweave-Underlay Hybrid Control Cannel Selection for Mobile Computing Services. In: *Second IEEE International Conference on Smart Systems and Inventive Technology (ICSSIT 2019)*, pp. 435–440. IEEE Press, Tirunelveli, India, (2019).

8. Manoharan, R., Balaji, S., Suthir, S., Obaidat, M., Suresh, K.C.:Selection of Intermediate Routes for Secure Data Communication Systems Using Graph Theory Application and Grey Wolf Optimization Algorithm in MANETs. *IET Networks*, 1–9 (2020).

9. Begam, S., Vimala, J.,Selvachandran, G.,Ngan, T.T.,Sharma, R.: Similarity Measure of Lattice Ordered Multi-Fuzzy Soft Sets Based on Set Theoretic Approach and Its Application in Decision Making. *Mathematics*, 8, 1255(2020).

10. Rajesh, M., Gnanasekar, J.M.: Path Observation Based Physical Routing Protocol for Wireless Ad Hoc Networks. *Wireless Personal Communications*97, 1267–1289 (2017). https://doi.org/10.1007/s11277-017-4565–9.

11. Rajesh, M.: Streamlining Radio Network Organizing Enlargement Towards Microcellular Frameworks. *Wireless Personal Communications* (2020). https://doi.org/10.1007/s11277-020-07336-9.

12. Vo, T., Sharma, R., Kumar, R., Son, L.H., Pham, B.T., Tien, B.D., Priyadarshini, I., Sarkar, M., Le, T.:Crime Rate Detection Using Social Media of Different Crime Locations and Twitter Part-of-speech Tagger with Brown Clustering,4287–4299(2020).

13. Nguyen, P.T.,Ha, D.H.,Avand, M.,Jaafari, A.,Nguyen, H.D.,Al-Ansari, N.,Van Phong, T.,Sharma, R.,Kumar, R.,Le, H.V.,Ho, L.S.,Prakash, I.,Pham, B.T.: Soft Computing Ensemble Models Based on Logistic Regression for Groundwater Potential Mapping. *Applied Science*, 10, 2469(2020).

14. Dansana, D., et al.: Using Susceptible-Exposed-Infectious-Recovered Model to Forecast Coronavirus Outbreak,*Computers, Materials & Continua*, 67(2), 1595–1612(2021).

15. Vo, M.T., Vo, A.H., Nguyen, T., Sharma, R.,Le, T.: Dealing with the Class Imbalance Problem in the Detection of Fake Job Descriptions.*Computers, Materials & Continua*, 68(1), 521–535(2021).

16. Sachan, S., Sharma, R., Sehgal, A.:Energy Efficient Scheme for Better Connectivity in Sustainable Mobile Wireless Sensor Networks.*Sustainable Computing: Informatics and Systems*, 30,100504(2021).

17. Malik, P.et al.: Industrial Internet of Things and its Applications in Industry 4.0: State-of the Art.*Computer Communication*,166, 125–139(2021),Elsevier.

18. Jha, S.et al.: Deep Learning Approach for Software Maintainability Metrics Prediction. *IEEE Access*, 7, 61840–61855(2019).

19. Analysis of Water Pollution Using Different Physico-Chemical Parameters: A Study of Yamuna River, *Frontiers in Environmental Science*(2020).https://doi.org/10.3389/fenvs.2020.581591.

20. Ghanem, S., et al.: Lane Detection under Artificial Colored Light in Tunnels and on Highways: An IoT-Based Framework for Smart City Infrastructure. *Complex & Intelligent Systems* (2021). https://doi.org/10.1007/s40747-021-00381-2.

21. Sharma, R., Kumar, R., Sharma, D.K., Son, L.H., Priyadarshini, I., Pham, B.T., Bui, D.T., Rai, S.:Inferring Air Pollution from Air Quality Index by Different Geographical Areas: Case Study in India. *Air Quality, Atmosphere, and Health*, 12, 1347–1357 (2019).

22. Sharma, S.et al.: Global Forecasting Confirmed and Fatal Cases of COVID-19 Outbreak Using Autoregressive Integrated Moving Average Model. *Frontiers in Public Health*(2020). https://doi.org/10.3389/fpubh.2020.580327.
23. Sharma, R.,Kumar, R.,Singh, P.K.,Raboaca, M.S.,Felseghi, R.-A.: A Systematic Study on the Analysis of the Emission of CO, CO_2 and HC for Four-Wheelers and Its Impact on the Sustainable Ecosystem. *Sustainability*, 12, 6707(2020).

7 Big Data Analysis and Machine Learning for Green Computing
Concepts and Applications

Monica Sneha, Arti Arya, and Pooja Agarwal
PES University

CONTENTS

7.1 INTRODUCTION TO BIG DATA

In today's technology-driven world, every individual every second is capable of generating 1.78 MB (MegaBytes) of data as per Jacquelyn Bulao [1]. With the advent of IoT (Internet of things) and social media platforms like Twitter, Facebook, YouTube and

DOI: 10.1201/9781003032328-7

WhatsApp, the data generation has become considerably easy. Though this data may seem unintelligible and inconsequential, they are of Business Importance. According to Facebook research [2], the user-generated data are analyzed to understand how Facebook users interact and improve the intercommunication amongst its users. As per Andrew Burst [3], while social media is one of the notable contributors of Big Data, public data and data generated by devices such as phones, IoT, and data warehouse applications also contribute to Big Data [4]. Thus, these bulks of data generated can be used to derive meaningful insights and also are of exceptional business value. These large volumes of data generated by us on a day-to-day basis are called Big Data.

Big Data can be defined as enormous amounts of data generated by various human activities that when investigated can contribute exceptional insights. The term "Big Data" was coined by O'Reilly Media in the early 21st century, addressing data that is huge and not easily manageable [5]. Big Data are not just enormous in volume but are also growing at a high rate, every second. The Big Data gathered from several sources are examined to identify and study patterns. The hidden insights gained can help in making decisions can help improve customer experience and can also serve as a prospective [6]. This Big Data, though can provide great insights, is far from manageable using traditional data analysis technique. Common methods adopted for analysis of Big Data are Machine Learning, Deep Learning and Data Mining [7]. Before the investigation, the pre-processing step of the Big Data is necessary to improve the performance of the analysis. Pre-processing improves the capability and effectiveness of Big Data analysis.

Big Data is not just a popular technology but is closely intervened with our day today living. One of the classic Big Data applications is the climate study, where immense climate data resources are investigated to interpret climate change. According to Hassani et al. [8], this climate forecast also serves other purposes such as agriculture, forestry, sustainable urban planning, disaster and disease estimation. In the current times, it is not just business using Big Data, but also the government sectors are using the same for approaching national challenges such as economy, education, poverty eradication and terrorism. Kim et al. [9] bring forth how various countries have invested in Big Data solutions for sustainable national growth and security. Big Data are capable of transforming our cities into a better habitable eco-system by addressing problems of traffic, safety, sustainability and other degrading problems prevailing in our current society. Allam and Dhunny [10] talk about how Big Data and AI (Artificial Intelligence) can collectively help us realize our vision of smart cities. Organizations can use Big Data analytics to understand how customers perceive their products and services, which will help them develop a better relationship with their customers [11]. Over the years, Big Data has also made its way into the lives of individuals through the OTT (over the top recommendation systems) platforms like the Netflix recommendation system. Maddodi and K [12] explain how Big Data analytics has enhanced the customer experience of Netflix and has been monumental in making it one of the leading and popular OTT platforms. Healthcare systems as explained by Bhuiyan et al. [13], operation and supply chain management as told by Addo-Tenkorang and Helo [14] and as per Yang et al. [15] for enhancing maritime research, are some of the other areas where Big Data has proved to be beneficial. According to Zhou et al. [16], startups that focused on Big Data-driven smart energy experienced a rapid growth.

7.2　THE EMERGING TECHNOLOGY: BIG DATA ANALYTICS

7.2.1　BIG DATA SOLUTIONS

Big Data has been rapidly growing, for a few years and with the advent of IoT and Man becoming more technologically dependent, Big Data is expected to expand more quickly. As the number of devices around us has increased, and with evolving technology over time, Big Data analytics is more on-demand subject now. According to Satyanarayana [17], Big Data analysis helps understand and target customers and also helps in understanding and optimizing business process [18]. Data monitoring and logging help businesses make versed decisions and help in improving customer experience. As data analysis helps in identifying bottlenecks, spot blind spots, making the business process efficient, businesses will also always choose to assemble data. The challenge lies in extracting useful information from this data, where Big Data analytics comes into play. Big Data has been leading businesses since prior years and now might take the lead [19].

7.2.2　THE 5VS OF BIG DATA

Big Data is characterized by three prominent features, namely, Volume, Velocity and Variety.

i. Volume

The indisputable feature of Big Data is its Volume. Big Data is all about volume and it is this property that has made it so valuable. These huge volumes of data have proved beneficial over time in various industries. It has been estimated that 2.5 Quintilian (10^{18}) bytes of data are generated every day [20]. While mobiles, Cloud Computing and IoT are considered the major drivers of Big Data, a swipe, search and share can generate data too. According to H. Arne [21], 59 zeta bytes of data is generated in 2020 and it is expected to rise to 149 zeta bytes by 2024.

ii. Velocity

Velocity defines the rate at which the data is growing every day. With the arrival of social media, the YouTube uploads, emails sent, Google searches, Instagram shares, Facebook likes, Online Shopping, Blogs, Websites and IP traffic all contribute to the velocity of data generation. Apart from these, the industry cloud data, monitoring systems and logs, add their part to the same. The Velocity of data generated requires us to process data faster to extract useful information from them [22].

iii. Variety

Since Big Data is generated from different sources, it is also in different formats. The data generated can be mails, pdfs, audio, video, posts and photos. Generally, this is classified as structured, unstructured and semi-structured data. While stock information, geolocation data are banking data are some examples of structured, emails, video recordings, log files and images are unstructured. The XML, JSON and.csv data files can all be considered as semi-structured. These different formats of data require different processing techniques, which poses a challenge in handling Big Data .

iv. Veracity

Veracity does not just define how valuable the data is, but also its quality and its authenticity. Veracity helps to identify what data is significant and what is not [23]. Veracity does not just validate data but also aims at improving it. Veracity involves many aspects of data, such as its source, quality and type. Veracity also identifies how to process data to the best of its ability to achieve our objectives [24].

v. Value

Value of Big Data involves estimating how useful the data acquired is. Processing Big Data requires high computational and storage resources; hence, it is necessary to evaluate the importance of the data. Understanding the value of data is important in making investment decisions and in availing the monetary benefits of the same. This attribute of data assesses the gains that can be derived from it [25].

Many other attributes are used to describe Big Data such as Variability, constantly changing and varying data as described by Analytics [26], Volatility and Validity. Khan et al. [27] also mentions two other Vs, Viscosity-Data Complexity and Viability-Data Activeness.

7.2.3   Stages of Big Data Lifecycle

Big Data is capable of providing rich insights, but we do not possess the ability to retain all of it. Hence it is necessary to identify the significance of data and plan on its holding. Regulatory policies and data storage costs determine the validity of the data, post which it is disposed of. Hence Big Data also involve a lifecycle where it is acquired from its source of generation, stored, analysed, retained or disposed of. Below we discuss the stages in the Big Data lifecycle [28].

7.2.3.1   Data Generation

According to Ghotkar and Rokde [22], the data-generating sources can be primarily classified into Human, Machine and Organization.

i. **Data generated by humans**: This data include the tweets, Instagram and Facebook posts, the YouTube videos, the Google searches, our comments on feeds and also our unofficial emails. Audio and video files have also come into picture due to the number of people posting media files on social media. These files create an immense amount of digital exhaust and this also includes the metadata. Metadata is the information about the files such as descriptive, structural or statistical information. While data takes up space, metadata uses more space. Another cause for data explosion is due to the existence of multiple replicas of the same data. Another challenge is that most of this data is unstructured and deriving insights requires more processing [29].

ii. **Machine-based generation**: The machine-generated data involve the logs generated by mechanical and digital devices. The IOT devices that are now a part of our lives hold sensors that generate data that are used to gain insights on customer data and also device behaviour. The control systems of industries generate logs that are useful in monitoring the systems. Health devices, IoT devices and sensors generate real-time data that is needed to be processed real-time to provide feedback for safe execution [30].

iii. **Organization-based generation**: The data generated by organizations are useful in deriving insights that can help improve businesses. This data is structured and usually is an agreed-upon format and are usually stored in relational databases. This data is used in operational and business intelligence system. Since this d is structured data, processing can be pre-planned and easier.

7.2.3.2   Data Acquisition

DataaAcquisition is the process of data collection, transport and pre-processing. Data acquisition firstly involves gathering data from distributed sources in different forms such as structured, unstructured and semi-structured forms. This data gathering is controlled by the 'V' of Big Data called Value. Often data from different sources are huge in Volume, high in Velocity and of great variety, but low in value. In such cases, it is necessary to adopt protocols that can increase the efficiency of the data gathering process. These protocols help in efficient gathering of data from distributed sources of distinct types. The protocols used are often based on the needs of the organization, but there are some standard protocols such as Advanced Message Queuing Protocol (AMQP). The AMPQ protocol assures ubiquity, data integrity and supports interoperability. A framework is used for data acquisition from distributed sources as guided by the protocol. The data thus retrieved are stored in persistent structures capable of analysis. The commonly used frameworks are Hadoop and Storm. According to S and Y [31], there are two types of data transportation, Inter-DCN (Data Center Networks) and Intra-DCN. Inter-DCN refers to the transport of data between the source and the data centre, whereas Intra-DCN is the movement of data within the data centre [30].

7.2.3.3   Data Storage

Big Data storage is facilitating a cost-effective and secure accommodation for the massive volumes of data so that it is available for analysis. Big Data storage is not only about storing large volumes of data but also about using a compliant structure to store data and to acquire intelligence from the same. There are some key requirements of Big Data storage systems and the most important of them all is their ability to scale. Scalable data storage is necessary to meet today's rapidly growing data and to be able to manage the incoming data rate. Another important factor is the input-output operations per second (IOPS) that the storage to provide to the analytical system [32].

- No-SQL

   No-SQL is an alternative to traditional relational databases used to take advantage of the fact that it can store a variety of data. No-SQL can hold documents, columnar models, key-values and graphs. No-SQL uses a flexible schema and uses the BASE (Basically Available, Soft state and Eventually consistent) Approach. Its capacity to store unstructured data and support agile analysis with sound performance delivers it beneficial for Big Data. Another useful property of No-SQL is that it scales horizontally which can be achieved by adding more commodity servers and storage, unlike RDBMS that scales vertically. MongoDB, Cassandra and HBase are some popularly used No-SQL databases for Big Data.

* NewSQL

    NewSQL databases are suitable for Big Data as they offer the SQL for Communication with the Databases and use the ACID (Atomicity, Consistency, Isolation, Durability) approach for Transactions. This database delivers the benefits of the NoSQL databases while retaining a few properties of the relational databases. NewSQL caters to the growing data in the OLTP (online transaction processing) systems and is capable of scaling without succumbing to bottlenecks. It also offers a non-locking concurrency control where the data read does not stall the data write that is performed concurrently [33].

* The distributed file system (DFS)

    DFS is the distribution of data across various servers and location. This enables users to obtain files that are physically distributed across storage systems using a common file access system. Distributed file systems facilitate the sharing of data amongst users in an orderly and authorized custom. DFS presents the transparency of data conveniently to users, where the user can access the data on servers as if it is on the client machine. Another advantage of DFS is its backup policy that maintains replicas of data, which serves the purpose of availability in case of data failure. The backup can be either a central access point or again a distributed access point, but the latter is preferred mostly as it makes the system fault-tolerant. The common examples of the Distributed Filesystems include HDFS (Hadoop Distributed file system), NFS (Network File System) and NetWare. The resources on the distributed systems can be accessed remotely, of which a few examples are Amazon S3, Microsoft Azure, OpenStack and Google cloud [34].

The storage technologies that are commonly used for Big Data are Direct Attached Storage (DAS), Network Attached Storage (NAS) and Storage Area Networks (SAN). DAS involves the disks present in the chassis of the server or could also be external disks connected to the server SAS card. DAS requires a smaller initial investment, but the scaling capability of DAS is limited, and it is susceptible to single-point failures. DAS is connected to a single server and works as dedicated storage for that server. Thus, it is difficult to share DAS across with another server [35].

DAS does not offer features of remote replication and snapshots. SAN involves the disks being connected to the virtual server via the fiber cables or iSCSI protocol. SAN is suitable for medium- to large-sized businesses as it requires a higher initial investment, but it can be configured to scale. NAS is a hybrid which uses a file-level transfer as opposed to the block level transfer used by DAS and SAN. NAS involves disks connected to the server using a TCP/IP connection. NAS is capable of scaling in terms of storage and also abstracts the underlying storage details from the server and also offers other benefits such as Thin provisioning, Snapshots and Replication.

The data stored can be queried using the Big Data query platforms like SparkSQL, Hive and Impala. Apache hive is used for managing data that is stored on the DFS using a Hive Query Language (HQL). Hive uses the traditional Map and reduces technique for data processing. Spark SQL is a structured data processing modules used to query structured data on a distributed system. Spark SQL has effectuated the Relational database querying style in distributed file systems. While it qualifies hive queries to be executed faster in the existing setup, it also holds a strong integration with the Spark ecosystem. Spark can process data sets and data frames and can be queried using Java, Python, Scala and R. In contrast to Hive, SparkSQL uses the structural query language and has made the Hadoop system more convenient. Impala is a massively parallel processing tool for analysis of data on the HDFS. It accepts the Hadoop file formats and preserves the Hadoop security. Its underlying framework is similar to that of Hadoop softwares and is capable of resource management is remarkable.

7.2.3.4 Big Data Analysis

I. Big Data Analysis Types

The Big Data analysis can use different analytical approaches based on the target data and the insight we are trying to derive. Below we discuss the different types of analytical techniques that can be used with Big Data.

i. Statistical analysis

Statistical analysis is the most commonly used approach for identifying trends and deriving insights from small datasets to massive data. This basically involves the use of numbers and formulas on data to infer conclusions. This approach is based on the statistical theory of mathematics, where uncertainty is resolved for using the probability. While descriptive statistical analysis is used to explain the data, inferential statistical analysis is used to conclude.

ii. Cluster analysis

Cluster analysis is used to group the objects under study. This is useful in identifying categories of objects that have similar properties and different from the objects in other groups. This is useful in grouping the objects of study and analysing the behavioural patterns in groups.

iii. Correlation analysis

Correlation analysis is used to identify relationships among the various factors under study. Mutual or correlated dependencies among the constituents of the study can be determined using this. This can be used to identify new correlation patterns amongst objects or use to prove or disprove hypothesis on existing correlation observations.

iv. Regression analysis

Regression analysis is used to identify the relationship between a target variable and the other variables under investigation. This is often used in cases where the correlation between variables is deceived due to randomness. This is used in complex problems to recognise the association of the factors involved in the study.

II. Big Data analysis design

Big Data analysis systems are designed to meet the needs of the business and based on the Volume and Velocity of the incoming data. Businesses may use real-time or offline analysis. In this section, we discuss the various designs adopted for data analysis.

- In today's world data is generated in a click, and there is a demand for extracting useful information from it instantly. To meet this requirement, real-time data analysis is vital. These handle incoming high-velocity data and gather insights in seconds. This is very useful in handling location data, in marketing and detecting anomalies. Real-time data analysis requires parallel processing clusters capable of handling high velocity and volume of incoming data. The Amazon Kinesis and Microsoft Azure stream analytics offer some of the data streaming tools for real-time analysis. When low response time is not required, offline analysis can be used. This is used when data is acquired into a specialized system where the analysis will be carried out. Here the system must be capable of acquiring and transmitting massive amounts of data to the analysis platform.

- The analysis design can also be further classified as In-Memory analysis, Business Intelligence analysis and Massive data analysis. The In-Memory analysis is suitable when the incoming data is smaller in size than the memory available for processing. In-Memory analysis can be performed real time and is useful if the entire data set can be housed in the RAM. In case of any changes in the data, the modified subset needs to be replaced in the memory before processing. Business Intelligence level is used when the data surpasses the size of the memory. In this case, the data acquired should be housed in a Business Intelligence platform, where it will be analyzed. This involves two parts, analytics and reporting, where complex computations are performed, and the results obtained are shared through the UI interface. Here the entire data are not available in the memory, and this impacts the performance negatively. When the size of the data exceeds the capacity of the BI level, the massive data analysis is used.

III. Big Data Analysis Tools

In this section, we discuss the various Big Data tools available for extracting knowledge from the available data. The generally used tools include Excel, R programming, Python and BI tools.

i. Excel

Though we use Excel in our day to day lives, we are not unaware of its great data analysis potential. Excel is capable of data processing and analytics, hence capable of data analytics for small and medium-sized businesses. Excel has plugins for data analysis that need to be enabled in order to be used.

Excel offers statistical features and visual properties to make data analysis possible. Excel also harbours the Analysis ToolPac that extends its features for statistical, financial and engineering analyses.

ii. R

R is an open-source programming language for data analysis and statistical computing. R has evolved from S programming language that was developed by AT&T Bell Labs for statistical analysis and visualization. R has developed from an editor to R studio, and now Jupyter Lab offers support to R. R is an open source and has grown in years due to the contribution of its user community. R offers Analysis, Statistics and visualization of data. R objects can be called in other Programming Languages such as C, C++ and FORTRAN and the functions developed in other languages can be called from R. R is a complete package for data analysis where we can clean data, perform dimension reduction, analysis, statistical modeling, statistical hypothesis testing, web crawling and data report output.

iii. Python

Python is a general-purpose programing language, whose one of the areas of application is Data Analysis. Python is superior to R concerning its data mining and machine learning program, except for the statistical capabilities of R. Python extends libraries for data processing, analysis, visualization and capable of faster processing. Python can also be used to write the Map-reduce programs of Hadoop, and it also provides PyDoop, a package to connect, read and write on HDFS. Python can resolve intricate problems using minimal programming efforts. In the current times, Python is exclusively used by data scientists as it grants a variety of libraries that make the data analysis process convenient.

iv. Business Intelligence

Business Intelligence tools are intended to speed up the business decision-making process by deriving useful insights from data within the required response time. Analysis of business data is intended to assist improve businesses achieve results, and using the appropriate tools can enhance the process and the results. The commonly used Business Intelligence tools are SAS, Matlab and SPSS.

7.2.4 BIG DATA PROCESSING METHODS

In this section, the common data analysis methods such as hashing, indexing and bloom filter are discussed below:

- Indexing

Data in the database is stored as records, and every record is identified uniquely by a key called index. Indexing provides faster and efficient access to large volumes of data. Indexes are generated based on the selected attributes of the databases, and they are of three, namely, primary, secondary and clusters. While primary indexing uses the primary or key attribute of the database to generate indices, secondary indexing uses the candidate key, and the cluster indexing is used for ordered data files.

- Hashing

  Hashing is an effective technique for addressing specific data blocks in massive volumes of data. Hashing helps in quicker reads and write by reducing the query time for data. Hashing uses functions called hashes to search for addresses of data blocks. Hashing is classified into static, which is used for data sets that do not grow dynamically and dynamic hashing, which is capable of locating data that dynamically grows or shrinks.

- Bloom Filter

  Bloom filter is a data structure used for rapid and efficient classifying of whether a data record is present in the data set or not. Bloom filter is a probabilistic space-efficient data structure designed to meet the need for faster searching of data records [36].

- Parallel Computing

  Parallel computing aims at processing data faster by splitting the task across multiple computing resources. In the case of Big Data, the task of locating the data record is split across computing resources to speed up the process. This is used by Map-reduce to locate data efficiently and faster using multiple levels of resources.

7.2.5 CHALLENGES OF BIG DATA

In this section, the challenges hurled by Big Data are discussed. The hurdles in the process of Big Data acquisition, storage and analysis are reviewed below:

  i. Big Data is the Gold Mine of Data, but it is just as good as its management and analysis process:

     Big Data has been a buzz word in the past decade and is expected to improve business decisions. Companies are collecting vast amounts of data and trying to derive insights to improve their business. Big Data is capable of all the hype around it, but only if handled, processed, and analysed competently. To prevent the Big Data analysis process from spiraling out of hand, it is necessary to design a Big Data management plan and process. It is also important to accept the learning curve and not hasten the results and focus on developing a better process.

  ii. Data Quality Complexity

     Since Big Data has been acquired from different sources in a variety of different formats, data integration of this data can pose a challenge. The incoming Big Data is not completely accurate and may hold a low data value. Incoming data may also be redundant and duplicate, and hence a data cleaning process is necessary. This issue can be faced by developing a data cleaning strategy and preprocessing the data to eliminate duplicates and redundant data. Redundancy elimination is achieved often by comparison with a single point of truth and using the match-merge technique [37].

  iii. The Upscaling Problem

     An undeniable feature of Big Data is its ability to grow, and this poses the scaling challenge. To be prepared for data explosion and to make upscaling easier, it is necessary to have a reasonable architecture that is designed keeping in mind the need to scale in the future. It is also necessary to keep in mind the scaling problem while designing Big Data algorithms. Also, it is necessary to have a maintenance and support team to attend to issues during the scaling process.

iv. Data Security

Data security is an important topic of discussion concerning Big Data. While Data security is an important factor for customer-oriented E-commerce businesses, this also plays a critical role in Healthcare devices, IoT devices and in banking transaction. The increasing number of cyberattacks emphasizes the need for focus on data security. This is a three-fold process where there are three important factors to be considered, namely, confidentiality, where the user data is kept private, second integrity, where the data accuracy and consistency are maintained and availability, where the data is accessible for processing. A solution to this aspect is to consider the data security problems in the initial levels of Big Data solution development and also in all stages of the Big Data process. It is time to stop ignoring data security and put it in the first place while designing the architecture. According to Zhang et al. [38], another important aspect of Big Data is its credibility. Since Big Data is acquired from different sources it is necessary to verify its credibility.

v. Variety of Big Data technology

In the era of Big Data, there are a variety of technologies available for handling and mining data. While it is an advantage to have many prospects on technology to meet our needs, it also poses a challenge of choosing the right technology. A few prominent technologies for Big Data analysis and management are HDFS, Apache Spark, Hadoop MapReduce, Kafka, Hbase, Cassandra, etc. With the availability of a variety of technologies, there is also a need to identify technology that suits the needs of the user. A few factors that determine the type of technology to be used include cost, data size, response time required, and the target results. To make the best use of the available technology, it is necessary to derive a strategy and use expert opinion to identify the right technology to meet your needs.

vi. Energy Management

Big Data requires massive storage and computing resource to store and process it. These computing resources require electric energy to function and, the large amounts of electricity needed raises financial and environmental concerns. The manufacture of computing resources, power consumption of them during the lifetime and their disposal all prove detrimental to the environment. To overcome these shortcomings green computing has been adopted by organizations.

vii. Big Data Workforce

Apart from the varieties of Big Data tools available today, the driving force of these tools, the Big Data professionals play an important role in the overall Big Data knowledge retrieving process. While there is a growing demand for data scientists, a lack of skilled professionals available poses a challenge [39]. According to Wani and Jabin [40], if training is provided for the Big Data professionals, a big difference can be made to the current scenario. Also, the hype around Big Data leads to unrealistic expectations by the organization that has caused the Big Data professionals to quit their jobs and in worst cases also made them change their career trajectory.

7.2.6 Technologies Involved with Big Data

In this section, we discuss the technologies that influence or are influenced by Big Data. These technologies counter impact each other and have been responsible for the rapid growth of Big Data.

I. Hadoop

Hadoop is an open-source framework for storing and processing Big Data. Hadoop has eased the processing of Big Data and provides features of fault tolerance and parallel processing possible for the same. Hadoop has transformed the data processing world by overcoming the limitations of speed and time. Hadoop has subdued the capacity issue by using distributed file systems and storing data records across, commodity servers that are economical and capable of scaling. Hadoop uses the Map-reduce paradigm to split the data to be processed across distributed servers so that it can be processed concurrently and returns the merged result from the servers. These processes reduce the time complexity of data processing in Hadoop. Hadoop is fault-tolerant as data stored on a node is replicated on other replica nodes, making it available in case of failures. The HDFS component Hadoop is capable of storing both semi-structured and unstructured data that can be parsed and fitted into a schema for processing, thus making Hadoop flexible for different data formats. Hadoop possesses the properties of resilience, data diversity, scalability, speed and low cost that make it an excellent choice for Big Data processing. The core components of the Hadoop system are the HDFS, YARN and the Map-reduce. HDFS consists of the name and data nodes that store the data records and this component for maintaining the distributed file systems and for replication. YARN (Yet Another Resource Negotiator) is a resource manager which schedules and manages the resources. The Map-reduce is the programming model responsible for fast indexing of data. Hadoop also offers other subsidiary components such as HBase, Hive, Kafka and Zookeeper. Hbase is a column-oriented database to resolve simple queries across a row with the low response time. It is a non-relational database and uses the hash tables to achieve low latency. Kafka offers faster data transfer using the publish-subscribe messaging system. Hive uses the HQL to query the Hadoop distributed system as an alternative to the complex Map-reduce queries [41].

II. IoT

Internet of things involves using sensors, software and other technology in devices to connect them to the internet for remote control and updates.

The IoT systems generate massive amounts of data, those captured by sensors and also the system logs. IoT has also been responsible for rapid growth in Big Data, due to the large volumes of data it generates. This data analyzed is capable of offering useful insights that can help improve the system itself. The real-time data generated by these sensors and logs are stored and analyzed using Big Data technology. Both IoT and Big Data are interdependent fields. While IoT was responsible for the rapid growth in Big Data, Big Data analysis offers insights and solutions to improve IoT systems.

III. Cloud Computing

Cloud computing is virtual processing of computational tasks on high powered servers. As opposed to limited computing resource availability in traditional computing, cloud computing provides an abstraction of better resource availability for the users. Cloud computing is capable of the more powerful computational requirements of Big Data. Cloud Computing can support and minimise Big Data analysis efforts.

7.2.7 FUTURE OF BIG DATA

Global data has been growing rapidly since the past decade, and similar trends are expected in the future. With exponentially ever-growing data, businesses and organization will try to extract useful knowledge and gain the best value of it. In a data-driven world, the Big Data market is expected to grow, and companies are expected to invest more in Big Data analysis technology. According to Reinsel et al. [42], by 2025, Big Data is expected to grow to 175 zettabytes and that 49% of the data will reside in the Cloud environment.

As Big Data continues to grow, machine learning will expand to impact Big Data to its full potential. Machine learning is developing along as it is used for business operations as well as other daily operations. With the advancement of open-source platforms, machine learning tools are available to the public for learning and improvement. Machine learning has found its way into several fields of application. Machine learning is now functional in several areas of application, speech recognition, computer vision applications, virtual assistants and recommendation engines. Machine learning has contributed to various fields like Healthcare, Automated Transportation, Robotics, Gaming, Banking and Digital Media.

As Big Data takes centre stage, data privacy and security will be a growing concern for businesses as well as individuals. Data confidentiality has always been a hurdle to Big Data, and with growing volumes, this can trigger greater confidentiality issues. With the constant evolution of cyberattacks posing a threat, the security skill gap is a step down towards data confidentiality. The only possible solution to this would be strict adherence towards data security standards, and constant updates are necessary to keep the business going. Thus the growing data have also provoked a positive growth in cybersecurity investments.

According to Khvoynitskaya [43], fast and actionable data will become the forefront in the coming years. The exponential growth in data demands a swift analysis of data to derive real-time insights to make fast business decisions. Processing of data in batch modes will be replaced with quicker real-time analysis to take immediate action on the incoming data. This can revolutionize how businesses work and provide more support to the agile working environments of businesses. It can also support various public sectors dependent on Big Data in making prompt actions in the time of need and uncertainty.

7.3 MACHINE LEARNING

Machine learning is a subdomain of Artificial Intelligence, which gives the computers an ability to learn without being explicitly programmed. Machine learning focuses on the properties and the performance of the learning systems. It is built upon the optimized schemes from many other fields such as statistics, information theory, mathematics, cognitive science and some disciplines of engineering. Machine learning has various application domains, a few of which are recommendation systems, healthcare, autonomous control systems, self-driving cars, malware filters and speech recognition. Machine learning is now prevalent in our daily lives, and some real-time examples where it is used are language translators and virtual assistants such as Alexa, medical diagnosis and also traffic prediction.

Machine learning is also used to explore, analyze and leverage Big Data. Machine learning is used in all phases of Big Data, starting from collection and analysis to integration [44]. Machine learning is capable of deriving insights from patterns found in Big Data and is likely to automate the decision-making process. Machine learning is used in Big Data operations such as data labeling, segmentation and scenario simulation.

Machine learning can be basically classified into three methods, namely, supervised learning, unsupervised learning and reinforcement learning, which we study in the below sections.

7.3.1 SUPERVISED LEARNING

Supervised learning requires labeled data set for training, where the data set has certain attributes along with the target attribute. Here the target attribute means the outcome or the target class. For supervised learning, the training data contain both the input attributes and output attributes. The output is predicted based on the labeled input data set. Supervised learning is further classified into:

i. Classification

Classification can be defined as the technique of predicting discrete class labels [45]. Classification models can be binary or multiclass. Binary classification involves predicting the inputs as one of the two target classes, whereas multiclass classification includes multiple target classes. An example of binary classification is the email spam filter which classifies emails into spam and non-spam. The classification of cyber-attacks into U2R (user to root) attack, DOS, R2U (Root to user) or DDoS attack is a good example of multiclass classification. K-Nearest neighbours, Logistic Regression, Decision tree and SVM (Support Vector Machine) are some of the commonly used classification algorithms.

ii. Regression

Regression is the technique of predicting Continuous Class labels. Regression consists of explanatory variables, which are independent variables that are the cause for the target variable, which is a dependent variable. Regression is further classified as simple regression and multilinear regression. Simple regression involves one independent variable, whereas multilinear regression consists of two or more independent variables. Simple regression is categorised into linear and non-linear. If the relationship between explanatory and the target variable is a linear function, then it is called linear else if it is a non-linear function, it is called non-linear regression. Prediction of the continuous daily opening and closing values of a stock is a good example of the regression model.

7.3.2 UNSUPERVISED LEARNING

Unsupervised learning, as its name suggests, does not require supervision. Here an unlabeled data set is used to train the model. Here the training data contain only the input variables and not the target or out variable. Unsupervised learning predicts the output based on the patterns in the input dataset. Dimension reduction, density estimation and clustering are the commonly used unsupervised learning methods [46].

i. Clustering

Clustering can be defined as the process of grouping similar objects, where every object is similar within the group and different concerning objects of the other group. A good example of clustering is the customer segmentation into platinum, diamond and gold classes. K-means, fuzzy C-means and DBSCAN have commonly used clustering algorithms. There are many real-world scenarios where clustering is used like exploratory data analysis, summary generation, fraud detection or noise removal, etc.

ii. Dimension Reduction

Dimension reduction, also called as feature extraction, is used to eliminate redundant attributes to render the prediction easier. Dimension reduction is the technique to overcome the curse of dimensionality of the training data set. Training data sets with very high dimensions can degrade the performance of the machine learning models; therefore, it is desirable to reduce the dimensions. Dimension reduction involves identifying relevant variables that impact the target variable while eliminating irrelevant attributes. The principal component analysis is one of the commonly used dimension reduction algorithms.

7.3.3 REINFORCEMENT LEARNING

Reinforcement learning is the process of learning by interaction with the environment where learning is based on the rewards it gains from the environment. Here the algorithm performs actions, and based on the impact of its actions, the environment offers it rewards. Based on the rewards from the environment, the algorithm learns to improve its actions. Reinforcement learning is used in scheduling elevators and robotic vacuum cleaners

7.4   BIG DATA, MACHINE LEARNING AND OUR ECOSYSTEM

Computing systems are a boon to mankind. They have changed our lives for ages and have influenced our lifestyle. The computing systems initially invented for calculations are now a part of many aspects of our daily life. Computers help us connect, communicate, learn, and be entertained, and for many, it provides a living. While computers have greatly influenced our lives, they also have drastically changed businesses and nurtured new ones. Many industries including Healthcare, Education, Marketing, Banking use computing devices, and every year, billions of systems are manufactured, used and are disposed. Though these systems have made our life easier, the process of manufacturing, the energy consumed by them and their reckless disposal have negatively impacted our ecosystem. Green computing aims at reducing the negative impact of these computing systems on our environment and strike a chord of balance between nature and technology. Green computing is a reasonable and responsible way of using computing resources so that we can minimize its adverse influence on the atmosphere. Green computing process involves being accountable for the resources used and careful disposal of the same. Though green computing aims to conserve our environment, it offers many other advantages in adopting this strategy. There are narratives of how green computing has benefited organizations and made them sustainable over time. Green computing is an ethical choice that has advanced many companies financially and has also contributed to customer satisfaction.

The objectives of green computing include energy efficiency, biodegradability and reusability of resources. El-Kassar and Singh's [47] study also suggests that green process innovation adds a competitive advantage for the organization and increases its performance. Hence, it is necessary to include a green strand in the emerging IT technology DNA. One of the most prominent growing technologies is Big Data Analytics, which has transformed knowledge discovery in the past few years. Hence, it is important to implant green practices in the Big Data Analytic Techniques which involve machine learning, data mining and statistics. According to Shuja et al. [2], doing so will be instrumental in using energy efficiently, increasing reusability and minimizing the computing and storage requirements of Big Data Analytics.

7.5   STATE-OF-THE-ART AND REAL-TIME
APPLICATIONS OF GREEN COMPUTING

In this section, we review the existing real-time green computing systems and the green practices adopted by the organizations. This is an attempt to understand the efforts required in implementing green practices and the outcome of the same in the real world. This section is dedicated for the case study of existing green computing systems in the real world.

In order to implement green computing on a large scale it is necessary to lay out a few practices that can walk us to our goal. Some of the common practices that have been followed are discussed here. Ojo et al. [48] explain a three-fold training approach that will be beneficial for anchoring the right mindset about green computing within the organization. The three steps towards green information technology is sharing green practical solutions, developing green management culture and lead by human capital. Anthony et al. [49] propose a life cycle strategy for green computing, where metrics are used to categorize data centre practice into measurable units that are useful in computing the cost-savings, energy efficiency, natural resources used and CO_2 emissions. According to Sourabh et al. [50], telecommuting also favours green computing as it reduces the fuel emissions caused by vehicles and also saves on the utility costs.

A classic flagbearer of the green Big Data technology is Google, the largest corporate buyer of renewable energy. Google is committed to reducing the carbon emissions of its compute resources to 0%. They strive towards efficient and smart data centres that are also de-carbonized and energy-efficient. Their first step towards this mission was of energy awareness, where the energy consumed by the server, storage, networking devices, cooling and lighting was measured. They further classified this energy as 'IT equipment energy' – that is consumed by servers, storage and network devices and 'Facility Overhead energy' – that is used for cooling and lighting [51]. The Power Usage Efficiency (PUE) is computed as the quotient of the total energy consumed divided by the Facility overhead energy. The PUE is computed on a quarterly and yearly basis to determine the amount of energy saved. The other practices that have helped them in this endeavour are optimizing airflow and raising the temperature to 27° as opposed to the popular belief that 22° were the optimal temperature for data centres, as per the revised recommendation of by the American Society of Heating, Refrigerating and Air Conditioning Engineers (ASHRAE) [52]. The green Big Data setup required an initial principal of 25,000 dollars, but this earned a savings of 670 MWh energy and 67,000 dollars of annual energy savings. Thus this leads us to conclude that adopting green practices benefits the environment as well as the one willing to commit to it. Though the initial work and investment involved may seem complex, its advantages over time are tremendous.

7.6   DISCUSSION

In this section, we would like to present some hidden benefits and indirect influences of green computing technologies. Though green practices require an initial investment in the long run, the gains are infinite. The green practices are meant to prevent the degradation of the environment due to the activities of humans, organizations and businesses. It aims at achieving energy efficiency, promoting reusability and recycling and biodegradability. Along with its contributions to the ecosystem, green computing has more to offer to us.

Over the years, businesses have demonstrated their social responsibility and have made an ethical choice to reduce their carbon footprints. In the above discussion, we have also witnessed organizations that have increased their profits by incorporating green practices. This noble choice does not just increase their revenue but also inflates their reputation amongst their customers and clients and illustrates their interest and the responsibility they hold. Green practices also impact the psychological wellness of their employees and boost their morale. It makes the employees confident about what they are striving to achieve and serves as a motivation to advance their endeavours.

The organizations that go green also enjoy tax benefits, which adds another reason to uphold the green virtue. The government offers a percentage reduction in taxes based on the green practices that the company has undertaken. An instance of these tax breaks is the 30% tax credit granted by the government for the use of renewable sources of energy such as solar and wind. Apart from solar and wind, fuel cells, geothermal systems and microturbines are all capable of tax returns. Green practices have also found to be impacting the stock markets.

The investors have shown interest in investing in environmentally responsible companies. To cater to progressive investors, certain Benchmarks, such as BSE-GREENEX in India, have been introduced. BSE-GREENEX is an index used to compare the performance in terms of their Carbon footprint. According to Bhattacharya [53], investment in green companies has rewarded the investors and does both the stakeholders and the shareholders are benefitted. Thus, we can conclude that green computing exudes a positive impact on various aspects though it was designed for a better ecosystem.

7.7 CONCLUSION

The chapter is intended to introduce Big Data and machine learning concepts to the reader while highlighting the importance of including green practices to make these technologies more effective. The topics and concepts covered by this chapter are listed below:

- The Big Data generation, acquisition, storage and the analysis techniques. The challenges, the state-of-the-art and the future of Big Data.
- The machine learning concepts and the correlation between Big Data and machine learning.
- Green Big Data technology and its benefits, and the current scenario of green computing

The chapter intends to emphasize on the need for green computing and helps the readers gain a perspective on the green practices that can revolutionize the technology world as well as our planet.

REFERENCES

1. J. Bulao, "How much data is created every day in 2020?" *tech jury*, 2020.
2. Facebook research, "Gaining insights to deliver meaningful social interactions", https://research.fb.com/category/data-science/, 2020.
3. A. Burst, "Top 10 categories for big data sources and mining technologies", *ZDNet*, 2020.
4. S. Begam, J. Vimala, G. Selvachandran, T. T. Ngan, and R. Sharma, "Similarity measure of lattice ordered multi-fuzzy soft sets based on set theoretic approach and its application in decision making", *Mathematics*, vol. 8, p. 1255, 2020.
5. M. van Rijmenam, "A short history of big data", *DataFloq*, 2020.
6. T. Vo, R. Sharma, R. Kumar, L. H. Son, B. T. Pham, B. D. Tien, I. Priyadarshini, M. Sarkar, and T. Le, "Crime rate detection using social media of different crime locations and twitter part-of-speech tagger with brown clustering", pp. 4287–4299, 2020.

7. P. T. Nguyen, D. H. Ha, M. Avand, A. Jaafari, H. D. Nguyen, N. Al-Ansari, T. Van Phong, R. Sharma, R. Kumar, H. V. Le, L. S. Ho, I. Prakash, and B. T. Pham, "Soft computing ensemble models based on logistic regression for groundwater potential mapping", *Applied Science*, vol. 10, p. 2469, 2020.

8. H. Hassani, X. Huang, and E. Silva, "Big data and climate change", *Big Data and Cognitive Computing*, vol. 3, no. 1, p. 12, Feb. 2019. [Online]. Available: http://dx.doi.org/10.3390/bdcc3010012.

9. G.-H. Kim, S. Trimi, and J.-H. Chung, "Big-data applications in the government sector", *Communications of the ACM*, vol. 57, no. 3, 2014. [Online]. Available: https://doi.org/10.1145/2500873.

10. Z. Allam and Z. A. Dhunny, "On big data, artificial intelligence and smart cities", *Cities*, vol. 89, pp. 80–91, 2019. [Online]. Available: http://www.sciencedirect.com/science/article/pii/S0264275118315968.

11. "How big data can help you do wonders in your business", *SimpliLearn*, 2020.

12. S. Maddodi and K. P. K, "Netflix bigdata analytics – The emergence of data driven recommendation", *International Journal of Case Studies in Business, IT, and Education (IJCSBE)*, vol. 3, no. 2, pp. 41–51, 2019.

13. M. A. R. Bhuiyan, M. R. Ullah, and A. K. Das, "ihealthcare: Predictive model analysis concerning big data applications for interactive healthcare systems", *Applied Sciences*, vol. 9, no. 16, p. 3365, Aug 2019. [Online]. Available: http://dx.doi.org/10.3390/app9163365.

14. R. Addo-Tenkorang and P. T. Helo, "Big data applications in operation/supply-chain management: A literature review", *Computers & Industrial Engineering*, vol. 101, pp. 528–543, 2016. [Online]. Available: http://www.sciencedirect.com/science/article/pii/S0360835216303631.

15. D. Yang, L. Wu, S. Wang, H. Jia, and K. X. Li, "How big data enriches maritime research – A critical review of automatic identification system (ais) data applications", *Transport Reviews*, vol. 39, no. 6, pp. 755–773, 2019. [Online]. Available: https://doi.org/10.1080/01441647.2019.1649315.

16. K. Zhou, C. Fu, and S. Yang, "Big data driven smart energy management: From big data to big insights", *Renewable and Sustainable Energy Reviews*, vol. 56, pp. 215–225, 2016. [Online]. Available: http://www.sciencedirect.com/science/article/pii/S1364032115013179.

17. L. Satyanarayana, "A survey on challenges and advantages in big data", 2015.

18. S. Jha et al., "Deep learning approach for software maintainability metrics prediction", *IEEE Access*, vol. 7, pp. 61840–61855, 2019.

19. R. Sharma, R. Kumar, D. K. Sharma, L. H. Son, I. Priyadarshini, B. T. Pham, D. T. Bui, and S. Rai, "Inferring air pollution from air quality index by different geographical areas: Case study in India", *Air Quality, Atmosphere, and Health*, vol. 12, pp. 1347–1357, 2019.

20. "Big data analytics industry report 2020- rapidly increasing volume & complexity of data, cloud-computing traffic, and adoption of IoT & AI are driving growth", *GlobeNewsWire*, March 2020.

21. H. Arne, "Volume of data/information created worldwide from 2010 to 2024", *Statista*, July 7 2020.

22. P. R. M. Ghotkar, "Big data: How it is generated and its importance", *IOSR Journal of Computer Engineering*, 2016, 01–05.

23. "Veracity: The most important v of big data", *GutCheck*, August 2019.

24. R. Sharma, R. Kumar, P. K. Singh, M. S. Raboaca, and R.-A. Felseghi, "A systematic study on the analysis of the emission of CO, CO_2 and HC for four-wheelers and its impact on the sustainable ecosystem", *Sustainability*, vol. 12, p. 6707, 2020.

25. S. Sharma et al, "Global forecasting confirmed and fatal cases of COVID-19 outbreak using autoregressive integrated moving average model", *Frontiers in Public Health*, 2020. https://doi.org/10.3389/fpubh.2020.580327.

26. S. Analytics, "Big data insights", *SaS Insights*.

27. N. Khan, M. Alsaqer, H. Shah, G. Badsha, A. A. Abbasi, and S. Salehian, "The 10 vs, issues and challenges of big data", in *Proceedings of the 2018 International Conference on Big Data and Education*, ser. ICBDE '18. New York, NY: Association for Computing Machinery, 2018, pp. 52–56. [Online]. Available: https://doi.org/10.1145/3206157.3206166.

28. P. Malik et al., "Industrial internet of things and its applications in industry 4.0: State-of the art", *Computer Communication*, vol. 166, pp. 125–139, 2021, Elsevier.

29. "Analysis of water pollution using different physico-chemical parameters: A study of Yamuna river", *Frontiers in Environmental Science*, 2020. https://doi.org/10.3389/fenvs.2020.581591

30. D. Dansana et al., "Using susceptible-exposed-infectious-recovered model to forecast coronavirus outbreak", *Computers, Materials & Continua*, vol. 67, no. 2, pp. 1595–1612, 2021.

31. C. M. M. S and L. Y, "Big data: A survey", *Mobile Networks and Applications*, vol. 19, pp. 171–209, 2014.

32. A. Adshead, "Big data storage: Defining big data and the type of storage it needs", *ComputerWeekly.com*.

33. M. T. Vo, A. H. Vo, T. Nguyen, R. Sharma, and T. Le, "Dealing with the class imbalance problem in the detection of fake job descriptions", *Computers, Materials & Continua*, vol. 68, no. 1, pp. 521–535, 2021.

34. S. Sachan, R. Sharma, and A. Sehgal, "Energy efficient scheme for better connectivity in sustainable mobile wireless sensor networks", *Sustainable Computing: Informatics and Systems*, vol. 30, p. 100504, 2021.

35. S. Ghanem et al., "Lane detection under artificial colored light in tunnels and on highways: An IoT-based framework for smart city infrastructure", *Complex & Intelligent Systems*, 2021. https://doi.org/10.1007/s40747-021-00381-2.

36. "Big data: From beginning to future", *International Journal of Information Management*, vol. 36, no. 6, Part B, pp. 1231–1247, 2016. [Online]. Available: http://www.sciencedirect.com/science/article/pii/S0268401216304753.

37. I. Lee, "Big data: Dimensions, evolution, impacts, and challenges", *Business Horizons*, vol. 60, no. 3, pp. 293–303, 2017. [Online]. Available: http://www.sciencedirect.com/science/article/pii/S0007681317300046.

38. D. Zhang, "Big data security and privacy protection," inProceedings ofthe 8th International Conference on Management and Computer Science(ICMCS 2018). Atlantis Press, 2018/10, pp. 275–278. [Online]. Available:https://doi.org/10.2991/icmcs-18.2018.56

39. A. Alharthi, V. Krotov, and M. Bowman, "Addressing barriers to big data", *Business Horizons*, vol. 60, no. 3, pp. 285–292, 2017. [Online]. Available: http://www.sciencedirect.com/science/article/pii/S0007681317300022.

40. M. A. Wani and S. Jabin, "Big data: Issues, challenges, and techniques in business intelligence", in *Big Data Analytics*, V. B. Aggarwal, V. Bhatnagar, and D. K. Mishra, Eds. Singapore: Springer, 2018, pp. 613–628.

41. A. Oussous, F.-Z. Benjelloun, A. Ait Lahcen, and S. Belfkih, "Big data technologies: A survey", *Journal of King Saud University – Computer and Information Sciences*, vol. 30, no. 4, pp. 431–448, 2018. [Online]. Available: http://www.sciencedirect.com/science/article/pii/S1319157817300034.

42. D. Reinsel, J. Gantz, and J. Rydning, "The digitization of the world from edge to core", *IDC White paper*.

43. S. Khvoynitskaya, "The future of big data: 5 predictions from experts for 2020–2025", *itransition*.

44. A. L'Heureux, K. Grolinger, H. F. Elyamany, and M. A. M. Capretz, "Machine learning with big data: Challenges and approaches", *IEEE Access*, vol. 5, pp. 7776–7797, 2017.

45. J. Qiu, Q. Wu, G. Ding, Y. Xu, and S. Feng, "A survey of machine learning for big data processing", *EURASIP Journal on Advances in Signal Processing,* Article no:67.

46. S. García, S. Ramírez-Gallego, J. Luengo, J. M. Benítez, and F. Herrera, "Big data preprocessing: Methods and prospects", *Big Data Analytics*, vol. 1, p. 9.

47. A.-N. El-Kassar and S. K. Singh, "Green innovation and organizational performance: The influence of big data and the moderating role of management commitment and hr practices", *Technological Forecasting and Social Change*, vol. 144, pp. 483–498, 2019. [Online]. Available: http://www.sciencedirect.com/science/article/pii/S00401625 17315226.

48. A. O. Ojo, M. Raman, and A. G. Downe, "Toward green computing practices: A Malaysian study of green belief and attitude among information technology professionals", *Journal of Cleaner Production*, vol. 224, pp. 246–255, 2019. [Online]. Available: http://www.sciencedirect.com/science/article/pii/S0959652619309461.

49. B. Anthony Jnr., M. Abdul Majid, and A. Romli, "A descriptive study towards green computing practice application for data centers in it based industries", *MATEC Web of Conferences*, vol. 150, p. 05048, 2018. [Online]. Available: https://doi.org/10.1051/matecconf/201815005048.

50. D. K. Sourabh, D. S. M. Aqib, and D. A. Elahi, "Sustainable green computing: Objectives and approaches", 2017.

51. Google, "Google's green data centers: network pop case study", https://static.googleusercontent.com/media/www.google.com/en//corporate/datacenter/dc-best-practices-google.pdf.

52. R. American Society of Heating and A. C. E. (ASHRAE), "Data center guide", *ASHRAETCS.*

53. R. Bhattacharya, "Article: Effect of going green on stock prices: A study on bse-greenex", *IJCA Proceedings on International conference on Green Computing and Technology*, vol. ICGCT, no. 1, pp. 32–37, October 2013, full text available.

8 Fundamental Concepts and Applications of Blockchain Technology

B. Rebecca Jeyavadhanam, M. Gracy,
and V. V. Ramalingam
SRMIST, KTR

Christopher Xavier
Barry-Wehmiller Design Group

CONTENTS

DOI: 10.1201/9781003032328-8

8.1   INTRODUCTION

The urge for up-gradation has pushed the rapid development of technologies in the past decade. Technologies such as Blockchain, Internet of Things, Machine Learning, Deep Learning, and Augmented Reality have become a vital part of day-to-day lives. Blockchain is a structure that stores transactional records known as blocks. Once stored, the record cannot be changed or deleted. The chain gets extended when new blocks are added at regular intervals and parallelly the size of the database also increased. Blockchain was made more famous when it was used to develop the most popular digital currency, Bitcoin, although many cryptocurrencies run on the blockchain. It revolves around the Internet; all on-line transactions in blockchain are well authenticated using cryptographic proofs and gaming theory consensus mechanisms. The consensus mechanisms already made blockchain near impossible to hack. Blockchain follows a peer-to-peer network that helps in information sharing between domains that do not trust each other [1]. The technology behind digital currencies is blockchain and we call the technology Distributed Ledger Technology (DLT). Although thousands of currencies are in existence, Bitcoin's concept is unique, and it now serves as a foundation for other cryptocurrencies. Blockchain utilizes cryptography for confirmation, to handle the information, and to check the exchanges in a record. Guaranteed certainty is there for the sections for being secure and liberated from misbehaviors. Cryptography is a memorable idea utilized for confirmation purposes and it was gained with the guide of blockchain to give a few wellbeing contributions like classification, security, and protection. Cryptography depends upon two straightforward components: a calculation and a key [2]. The calculation is a numerical count and the key is the significant segment utilized in the estimation. Blockchain can be thought of as a decentralized public ledger that keeps track of all committed transactions in a sequence of blocks. When new blocks were added to the chain, it grew longer. Bitcoin is the most valuable digital currency by market capitalization.

The main characteristics of blockchain are as follows:

- Transparency
  Each node in the blockchain is transparent and every single bit of data can be traced right back to its original position.
- Immutability
  Data stored inside is cannot be meddled with since the blockchain is secured using cryptographic hash functions.
- Decentralization
  Without any third-party trust, the blockchain concept is an essential part of this decentralization together with the process of authentication and authorization with hash functions, digital signature, and distributed consensus mechanisms.

Blockchain technology gained its momentum when the cryptocurrency Bitcoin invented and helps in solving the double-spending problem. In a P2P network, any peer can perform a transaction. Blockchain makes use of a consensus approach to get to the bottom of these sorts of conflicts. Figure 8.1 gives more clarity about the Client/Server network and Peer-to-Peer network [3].

Client/Server systems are extensively used in initiative applications. It is a computing system where the server computer achieves all resources and client computers are associated with the server. It follows the request/response model and its services are circulated over the network.

A **Peer-to-Peer** (P2P) network is a distributed construction that is widely used in applications that ease online request transfer such as content distribution and file sharing. In this figuring model, every computer/participant in the network also called peers is equally honored. This architecture is a decentralized network and does not support the Client/Server technology [4].

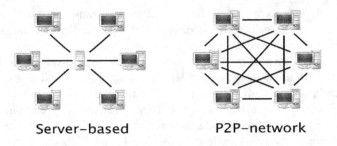

Server-based   P2P-network

FIGURE 8.1   Server-based and peer-to-peer network.

8.2   BLOCKCHAIN ARCHITECTURE

Transactions in the blockchain were done in a sequence of blocks. All participants in the network and ledger will be visible to everyone. A peer-to-peer network is used for the synchronization of the work. The two cryptographic mechanisms used by blockchain are hashing and digital signature. These mechanisms will be explained later in this chapter. Figure 8.2 illustrates the structure of a blockchain. In a typical blockchain, a block consists of main data (transaction records) and transactions in the blockchain are encrypted and coordinated between the nodes. Pointers are used to point the position of the next block. Each block points to the immediately previous block with a reference to a hash value of the previous block. Each past block is the parent of the resulting block. 'Genesis' is the name given to the first block of the blockchain network. A link to the previous block is generated by a secure hash algorithm and the hash value will generally be in 256 bits. Timestamp denotes the time at which any block is created. Nonce offers a fixed-size string that can consist of mathematical and alphanumeric characters [5]. Blockchain uses a unique function called Merkle tree (Merkle tree root) to create an ultimate hash value. The hash of the Merkle root can be used as an absolute mechanism to verify the authenticity of a block. Even a small variation to any of the contents in this tree will alter the value of the unique Merkle root. Merkle root helps in reducing data broadcast time. The Merkle tree root is shown in Figure 8.3 [6].

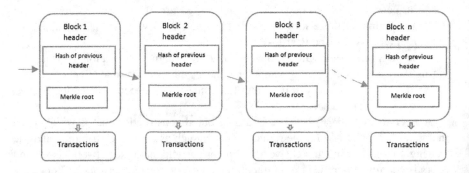

FIGURE 8.2   Structure of a blockchain.

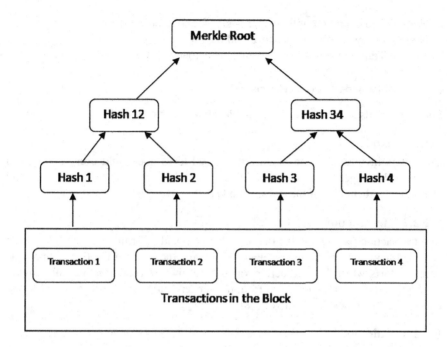

FIGURE 8.3 Merkle tree root.

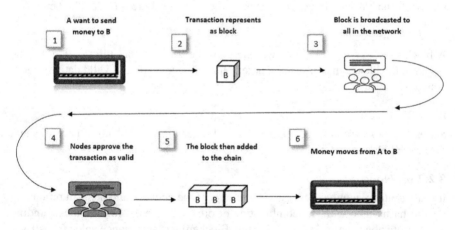

FIGURE 8.4 Working process of blockchain.

8.2.1 BLOCKCHAIN WORKING

The working of blockchain is shown in Figure 8.4, and the sequential process is shown in the following steps:

Step 1: A node requests to make a transaction (A wants to send money to B).
Step 2: The transactional operation is denoted as a genuine block.
Step 3: The transaction request is broadcasted to all other blocks in the network.

Step 4: The operation will be approved if found to be valid.

Step 5: A new block is added after validation.

Step 6: Transaction marked as complete. (Money moves from A to B.)

8.2.2   KEYWORDS USED IN BLOCKCHAIN

Some keywords are repeatedly used in blockchain technology. They are as follows.

8.2.2.1   Node

Computers involved in blockchain are referred to as nodes and every node has a replica of the ledger. Every node/participant will take part in all the transactions and checks the authenticity of the transaction [7].

8.2.2.2   Transaction

The transaction part maintains the records, data, and information. The transaction area will have sender details, receiver details, and the value of the transaction. The process starts when the transaction value is received. Cryptographic tools, digital signature, and hash are generated from the sender's side, and the transaction is dispersed over the blockchain network [8].

8.2.2.3   Address

Addresses are the unique identifiers for representing the sender's and receiver's information of the blockchain network. During the start, an address is a public key.

8.2.2.4   Block

A blockchain is a digital database, where all the transactional information will be stored in the blocks in consecutive order. Block is a collection of valid data and generated by the hash algorithm.

8.2.2.5   Chain

Series of blocks are arranged in a sequential order to form a chain of networks and thus called blockchain.

8.2.2.6   Nonce

It is an arbitrary number generated by miners of the network and helps in producing a valid hash value with the combination of other parameters in the block. Finding appropriate nonce value is called Mining. Block miners start nonce value from 0 and increase the value by 1 until a valid hash is found. The difficulty increases when the number of valid hash values decreases. The mining process is not a simple task as it takes more computational power for finding a valid hash [9].

8.2.2.7   Mining

Mining involves the formation of the hash value for the block of transactions. The transaction should be in such a way that no one can attempt to do any malpractice. Thus, the integrity of the chain is assured. Persons who involve in mining are called miners. Miners are the persons who run the blockchain and can add a block to the blockchain. Several miners can compete to create a block in the blockchain and all mining will be suitably rewarded [10].

8.2.2.8 Consensus

The consensus is defined as a set of rules to be surveyed for adding a new block into the structure. Certain consensus mechanisms are there to process before a block gets added. Consensus mechanisms are explained in detail in the subsequent pages.

8.2.2.9 Hard Fork

A hard fork is a rule change for authenticating the block. All nodes should work in agreement with the set of rules. A fork is formed in a scenario when two miners are finding the block in parallel [11].

8.2.2.10 Merkle Tree

Merkle tree is used to make a single or final hash value for a block. The closing hash value is also called a Merkle root. Merkle tree root allows effective and protected authentication of the contents of huge data structures. More usage of computing resources can be reduced.

8.3 TYPES OF BLOCKCHAIN

Current blockchain systems can be categorized into three types and represented in Figure 8.5.

8.3.1 Public Blockchain

Public blockchains are termed as a permission-less blockchain with no limitations on participation. Any public blockchain is a decentralized system open to everyone and where the distributed ledger is updated by numerous unsigned users. No one can act as a boss which means that nobody will have whole control over the network. Hence, it guarantees that the data is sheltered and helps in the immutability of the records [12].

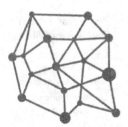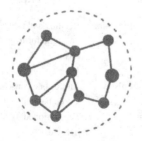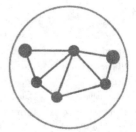

FIGURE 8.5 Public, consortium, private blockchains.

8.3.2 PRIVATE BLOCKCHAIN

Private blockchains are termed as permissioned blockchain with all transactions are visible only to persons who are part of the blockchain network. Private blockchains are centralized and are administered and controlled by someone who can make sure that the authorities are guiding participants. It can be a bank or an institute. Permissioned blockchains are limited to valid users of the network. All transactions are encrypted by a private key and cannot be interpreted by any person. These block- chains are typically used in private organizations like hospitals and universities to store sensitive information about them [13].

8.3.3 CONSORTIUM BLOCKCHAIN

The combination of public and private blockchains makes a way for another type called consortium blockchain. In this type, some nodes are private, and other nodes are public. Thus, some of the nodes will be permitted to participate in the transac- tions and some are to switch the consensus process. Consortium blockchains can be called hybrid blockchains because they can either be private or public blockchain depending on the situation [14].

Table 8.1 shows the comparison of public, private, consortium blockchains with different perspectives.

8.4 TIERS OF BLOCKCHAIN TECHNOLOGY

Because of the quick turn of events and progress made in blockchain innovation, numerous applications will get progressed over the long. Some applications have already been apprehended while some can be intended for the future based on the current pace of development in blockchain technology. The tiers of blockchain are characterized based on applications in each category [15].

TABLE 8.1

Comparison of Public, Private, and Consortium Blockchains

Perspectives	Public	Private	Consortium
Access	Reading and writing – for a single form	Reading and writing – for a single form	Reading and writing for multiple selected forms
Speed	Slower	Faster	Faster
Efficiency	Low	High	High
Security	Consensus mechanisms	Multi-node consensus	Multi-node consensus
Immutable	Data cannot be hindered	Could be hindered	Could be hindered
Consensus process	Permission-less	Permissioned	Permissioned
Participant's risk	Malicious	Trusted	Trusted

8.4.1   BLOCKCHAIN 1.0: CURRENCY

Blockchain 1.0 was introduced with the invention of Bitcoin and is used for crypto-currencies. The version allows monetary dealings based on blockchain technology or DLT (Distributed Ledger Technology).

8.4.2   BLOCKCHAIN 2.0: SMART CONTRACTS

Blockchain 2.0 can be called *Smart Contracts*. They consist of independent computer coding that executes automatically, and their conditions are defined well in advance. It is difficult to alter or work Smart Contract to decrease the expense of confirmation, execution, arrangement, and extortion prevention. **Ethereum** blockchain is the most prominent blockchain, which allows the implementation of Smart Contracts [16].

8.4.3   BLOCKCHAIN 3.0: DAPPS

A decentralized application (DApps) uses decentralized storage and communication and has its backend code running on a decentralized peer-to-peer network. On the contrary, traditional applications have their backend code running on centralized servers with centralized infrastructure. The main characteristic of a DApp is that it can have frontend code and user interfaces typed in any language and, like a traditional application, make calls to its backend [17].

8.4.4   BLOCKCHAIN 4.0: INDUSTRY 4.0

Blockchain 4.0 is used in the industry as it demands. Automation, resource planning, and a combination of different execution systems are some of the business demands. Blockchain 4.0 defines methods and solutions to provide the needs of business demands. Blockchain also gives a high level of trust and privacy protection [18].

8.5   BACKGROUND OF BLOCKCHAIN

Blockchain technology is built on the groundwork of distributed computing, software engineering, cryptography, and game theory.

8.5.1   BITCOIN

Bitcoin is the first-ever cryptocurrency with global attention. Bitcoin impacts blockchain and made the technology very popular. Bitcoin is not convertible and has no physical form. All computers in the blockchain network participate in the race of mining tokens. The race here denotes computational race. The objective is to come up with a random answer to the math algorithm. Since the algorithm is so complex, the only way to get the desired result is to make several predictions. More guesses from the computer will lead to winning the puzzle. Bitcoin has no governing body to set the rules and regulations.

8.5.2   DEVELOPMENT OF BITCOIN

Bitcoin's notion came into existence with the Bitcoin white paper authored by Satoshi
Nakamoto, whose identity is still not known to the world. Bitcoin cryptocurrency has
come a long way over the years amidst controversies and successes. The turning point
happened with the publication of a white paper by Satoshi Nakamoto, 'Bitcoin – A peer-
to-peer electronic cash system,' on October 31, 2008. Nakamoto himself successfully
created the Genesis Block (the very first block with no parent block) of the Bitcoin blockchain on January 3, 2009. By then the cryptocurrency assignment got a new dimension which has left an incredible mark across many industries. The first-ever Bitcoin
transaction of 10 BTC took place on January 12, 2009, between Nakamoto and the late
Hal Finney. The first-ever Bitcoin exchange rate was set against the dollar on October
5, 2009. At that time the equation point calculated as \$1 equaled 2300.03 BTC. A big
milestone happened on December 11, 2014, when Microsoft began accepting Bitcoin
payments. This made a path for companies across the globe to approve this vision.

8.5.3   WORKING PROCESS OF BITCOIN

A Bitcoin client or Wallet application is a program that helps the user to join the
Bitcoin network. A transaction is a process to control the drive of Bitcoin.

Step 1: Transaction starts by the user through the wallet application to transfer
coins from the sender to the receiver.

Step 2: Transaction invite is immediately broadcasted to all the miners (participants) in the blockchain network. The transaction will wait until the
miner chooses the one. If no one takes the transaction, it will be called an
'unconfirmed transaction'.

Step 3: Miners select the transaction from the pool of transactions, where
every transaction will be waiting to be chosen. A miner can create a block
of transactions of his own. However, there is a chance for multiple miners to
select the same transaction. When this type of situation arises, miners will
set priority to transactions based on the fee set for a particular transaction.
It is obvious that if the transaction fee paid is higher, the preference will be
given to that highly paid transaction.

Step 4: The first signed block is appended to the chain of blocks. This will be
done once the miner solves the complex mathematical puzzle. This process
of solving the puzzle is called Mining. Solving the puzzle needs more electrical energy for computational usage and extraordinary talent.

Step 5: Mining is a phase termed as a consensus algorithm in the Bitcoin
blockchain network. Proof of Work (PoW) is the name of this consensus
mechanism. The solved value by the miner is called a hash value, which will
start with a positive number of consecutive zeros.

8.5.4   POTENTIAL RISKS IN BITCOIN

In the contemporary world, Bitcoin is the most successful cryptocurrency. There are
some obstacles too when the attention turns to investment. Many people rushing to

invest in Bitcoin and awareness are needed to know about the concerns surrounding this new souk.

8.5.4.1 Volatile

The estimation of Bitcoin is consistently evolving. The Bitcoin market is continually moving every so often. With a particularly unpredictable market, it is consistently a danger to contribute. To evade a tremendous misfortune, watch out for the market and make the supports little.

8.5.4.2 Cybertheft

Digital currency is innovation-based, which lets Bitcoin reserve funds open to cyber-attacks. Hacking is a genuine danger, and whenever hacked it is almost difficult to recover the lost or taken Bitcoins.

8.5.4.3 Fraud

Customers and retailers are hoping to trade Bitcoins on the web, yet the expanding acknowledgment of this exchange can make trades counterfeit. The Consumer Finance Protection Bureau (CFPB) and the Securities and Exchange Commission (SEC) have advised against counterfeit exchanges and financial backers are cheated out of their Bitcoins in counterfeit trades. A decentralized framework is available to all and high danger is there with these sorts of security issues.

8.5.4.4 Technology Reliance

Bitcoin is an online trade that vigorously depends on innovation. Bitcoins are carefully mined, traded through keen wallets. Without technology, cryptocurrency is of no worth as it has no physical collateral to support it. People invest in gold, real estate, or bonds as these are physical commodities to exchange and digital currency is fully technology-based and easily vulnerable to cyber threats.

8.6 CONSENSUS MECHANISMS

It is consensus mechanisms that help blocks to be added into the structure. In the blockchain network, all nodes should reach a common agreement for adding a new block. The consensus mechanisms will help nodes to come to a common agreement. This confirms the trust between unfamiliar peers of the network. There are many consensus mechanisms and we will discuss few mechanisms briefly.

8.6.1 PROOF OF WORK

PoW is an initial consensus mechanism developed and used in the Bitcoin cryptocurrency process. This algorithm is extensively used in the mining process. Miners are attempting to solve the computational puzzle to add a block that is new to the network. A computational puzzle is nothing but to obtain the hash value for the next block and it needs high computational power and the process is a time-consuming one. When a miner obtains the relevant hash value, it will be broadcasted to all in the chain. About 50% of nodes should approve the hash value as correct. After this, the block will be added to the network and the miner will get Bitcoins as the reward for solving the puzzle.

8.6.2 PROOF OF STAKE

Proof of work mechanism needs more electrical power and results in high energy consumption. Alternate to PoW is Proof of Stake (PoS). PoS algorithm authenticates the block according to the stake of participants and uses a pseudo-random selection process to pick a participant to be the authenticator of the next block, based on the amalgamation of staking age, randomization, and the participant's wealth. Blocks are forged rather than mine. Persons who want to participate in the forging process are required to invest a certain number of coins into the network as their stake. The bigger the chances are if the stake is big.

8.6.3 PRACTICAL BYZANTINE FAULT TOLERANCE

Byzantine fault is the situation when different onlookers get different indications. The failure is any kind of system service damage due to a byzantine fault. So Practical Byzantine Fault Tolerance (PBFT) needs every node to be known to all in the network. The PBFT can accomplish up to 1/3 hateful byzantine copies. When a new block is determined, the process can be categorized into three different stages like pre-prepared, prepared, and commit. Each round needs to select a key, based on the consensus guidelines and it would be accountable for meeting the operation. If a node has to enter the next stage, it needs to obtain votes from 2/3 of all nodes.

8.6.4 PROOF OF BURN (POB)

To affluence, the loss of PoW, Proof of Burn (PoB), was introduced to miners. It is an alternative to the PoW consensus algorithm. Proof of Burn appeals to miners to send their coins to address where they cannot be elated. By burning or destroying Bitcoins, miners get an opportunity to mining. The coin burns reduce the circulating source. Proof of Burn inspires long term promise by the miners and PoB avoids the double-spending problems.

8.7 APPLICATIONS OF BLOCKCHAIN TECHNOLOGY

Blockchain technology is getting commercialized, and numerous industry clusters are coming out with use cases showing that the technology could be proven suitable for different communities. Abundant blockchain applications and platforms are broadly open, starting with Bitcoin, trailed by Ethereum, which goes about as a stage for developing decentralized applications utilizing brilliant agreements. Ethereum utilizes another idea of a symbolic economy, though Bitcoin utilizes a coin economy. Developing applications of Blockchain technology are in voting, digital identity, banking, supply chain, and health sector. All these applications demonstrate how blockchain can hypothetically be used to address comprehensive business challenges.

8.7.1 FINANCE

Blockchain applications have been applied to financial services in numerous ways and getting plentiful benefits. Smart contract service supports in directing financial transactions without an intermediate. It has the potential to accomplish securities, actions, payments, and privileges in a computerized manner. Banks and other financial institutions are started using blockchain.

8.7.2 INSURANCE SECTOR

Blockchain technology is bringing a revolution in the insurance segment by fetching optimization incorporate developments and sharing the information with improved competence, security, and transparency. It is bringing the strategy shift into the insurance system from physical to computerized using the smart contracts on the P2P networks and thereby removing the old-style processing system. There are numerous reimbursements where insurance firms and individuals looking for insurance and blockchain technology can meet the requirements. With pure decentralization, the insurance segment will get efficient in endorsing, expenses, privileges, and additional coverage processes. High security is guaranteed as the data transactions will be done in a decentralized fashion instead of a centralized way, where it is no longer under single entity control and thereby provides protection to the data and saving the system from spending more money.

8.7.2.1 Benefits Offered by Blockchain Technology in Insurance

Blockchain offers benefits to overcome the flaws in the insurance sector.

Health insurance
> Blockchain technology improves health insurance and also renovates the services of healthcare workers. Health insurance is in direct link with the medical organizations and patients will put to advanced data analytics for all diagnostic procedures. Such procedures and actions can be done through distributed ledger technology concepts in an efficient, safe, immutable, and transparent manner.

Auto insurance
> The auto insurance industry can profit in terms of reducing the level of form-filling, data repository related to previous maintenances, and compensations to a vehicle in an immutable, transparent, and distributed style. This paves way for a quicker resolution to all types of claims mainly accident claims.

Life insurance
> A lot of development is essential in the prevailing life insurance system in terms of removal of excess paperwork, the need for transparency in death claims genuineness, and funds transfer to heirs. With the help of smart contract technology, insurance firms can systematize the whole processing smoothly with reduced time and money.

Travel insurance

Travel insurance uses blockchain technology to protect the traveler from all his grievances such as flight delays, cancellation of flights, without checking with the airline office often. Operational efficiency in travel insurance will lead to smart international coverage, smooth travel across the globe.

8.7.3 MUSIC INDUSTRY

Blockchain technology has changed the music industry and allowed musicians to raise further. This technology helps in processing ownership rights of the music and providing reasonable payment to the musicians for their work in a transparent method.

8.7.3.1 The Glitches with the Present Structure of the Music Industry

The major problems with the present structure in the music industry are the nonexistence of transparency, clarity in rights, distribution of royalty, and the scuffle to monetize music files digitally. Reliability and accuracy are vital to safeguard that music makers and owners get proper payment for the work done. Problems that arise music industry are royalty distribution, copyright issues, patent rights, and payment issues. Blockchain eases out things by using smart contracts and provides a reliable and precise database of music files, music rights, the royalty of co-writers, partners involved, music producers, and give transparency in the system with prompt distribution.

8.7.4 IDENTITY MANAGEMENT

Blockchain technology shows its involvement in transforming identity management across the globe. The work done by identity management is tracking and dealing the digital identities in a protected and well-organized manner. These result in a decrease in data leak and swindle. Identity verification and authentication are essential aspects of every industry.

8.7.4.1 Difficulties in the Existing Identity Management System

There is no dedicated platform for identity management; people involved in this need to verify and authenticate their identities to whatever service they are assigned to gain. Recent technology has brought biometric identification which is password-based and stores all information in an unsafe structure which is highly disposed to data theft and hacking. The identity-centric centralized systems are using the concept of getting complete details of a person and this information is put under a unique social security number. The leak of a unique number and its code will result in disastrous acts like frauds in banking, procuring, blackmailing, and false identity creation. Some companies even trade people's personal information for commercial needs and generate money. The blockchain's decentralized mechanism of identity management benefits to resolve all such issues by providing a new model of identity management using Blockchain technology set-up. Cryptography is the concept of blockchain which helps in separating data from the identity of entities for safekeeping. This will help the company or individuals can preserve their privacy.

8.7.5   SUPPLY CHAIN

When the manufacturing/business process is taking place across the globe, the necessity of transparency between suppliers and supply chains is essential. One of the applications of blockchain is in supply chain management. It needs real-time tracking of goods and particularly pleasing to companies having various supply chains. With the help of Blockchain technology, all ineffective and unskilled supply chains will be removed. Trades are getting altered with the help of blockchain-based supply chain solutions. It offers endwise decentralized procedures through distributed ledger technology and digital public ledger. The supply chain is clubbed with the logistics industry, cargo, trucking, shipping, and all available modes of transportation which we use to transport goods. The expected need is to streamline the supply chain process and make it transparent.

8.7.5.1   Problems in the Old Supply Chain Management

The existing supply chain management system is outdated and unable to deliver solutions according to the pace the global economy is moving. Major problems facing the supply chain are the true value of the transaction, high cost, lack of transparency, and unproductive systems. The swiftness of the existing supply chain is extremely slow, the risk of faking and fraud always happens, deficiency of trust, unreliability, and uncertainty in data are some problems that have been detected.

8.7.5.2   Benefits of Supply Chain on Decentralization Platform

There are numerous benefits to the decentralization of the supply chain. The major benefits of a decentralization supply chain are its transparency, traceability, and immutability. These can be sensed in the real-time process. Real-time tracking of data is easily possible, and it helps in locating items with their condition and progressively reducing human fault. Transaction speed and productivity levels also improved. Malpractices are easily caught up in blockchain technology's trustless chain. Security is attained in all aspects. Benefits include improved account management, lesser courier costs, less correspondence, faster problem identification, satisfied customers, and more time to invent better products. Appropriate execution of the distributed ledger can be effective for the pharmacy sector to maintain the chain of protection over every drug.

8.8   CHALLENGES IN BLOCKCHAIN TECHNOLOGY

Since technology is growing rapidly, challenges are also raised at the same pace. Few challenges to the list are discussed as follows.

8.8.1   SCALABILITY

Bitcoin's block size is 1 MB fixed size. The size of the block has become a triggering concern for delays in transactions. The consensus of blockchain is any block being added to the blockchain is not allowed to weigh more than four million. Bitcoin blockchain was in its initial phase of growth and the network decentralization was confirmed by allowing low bandwidth nodes to a fixed size to join the network. The constraint on block size was introduced at this point. But with the current size of the Bitcoin network, transactions increased a greater number of times and one needs to look for solutions to tackle the scalability issues. For approval reason, all interchanges showing up in the blockchain must be put away. The limitation of block size and the period used to make another block has made the Bitcoin blockchain deal close by to ten transactions each second. This measure of exchanges cannot achieve the need of preparing a great many exchanges in the current world. Another issue is the size of blocks is paltry and little exchanges are getting deferred as miners incline toward high expense exchanges.

Scalability can be addressed in two ways, storage optimization and redesigning the block size. Blockchain must be reshaped by regulating block size and optimizing the storage.

8.8.1.1 Improving the Storage of Blockchain

The scalability problem can be solved in the blockchain by an innovative cryptocurrency outline proposed by Bruce in three categories.

- **Mini-blockchain**: This allows the latest portion of the blockchain to synchronize with the network.
- **Proof chain**: Cost of safety during the organization process can be resolved using a proof chain.
- **Account tree**: A data set that holds the equilibrium of all non-void locations.

8.8.1.2 Redesigning Blockchain

For upgrading blockchain, a person named Eyal proposed a novel thought called Bitcoin-Next Generation. The primary impression is to separate the customary block into two sections, one for selecting the transaction and other for storing the transactions.

Miners are contending themselves to turn into a leader to assume full liability for the micro block. The miner who turns into a leader will lead the participants gathering until another leader shows up. Bitcoin-Next Generation likewise extended the lengthiest chain approach where weightage will be on key blocks and micro blocks convey no weight. This is the way the blockchain is reshaped and the changes in the size of the block and capacity streamlining have been tended to.

8.8.2 PRIVACY LEAKAGE

The blockchain is intended to be an ensured situation as controllers just make exchanges with created addresses. There are chances where clients produce numerous locations. One instance of creating more locations is from data leakage. It was by a person named Kosba, who revealed that blockchain cannot maintain its privacy in transactions because all transaction details are publicly visible using public key. The recent learning shows Bitcoin transactions of a particular user can be linked in such a way it can be revealed. For this problem the suggested method is to link user pseudonyms to IP addresses, to get rid of easy revealing. By this, each user can be recognized by a set of nodes it links to. Several approaches have been proposed to progress the anonymity of blockchain. Anonymity denotes the non-identifiability of the sender and the receiver in one transaction.

8.8.3 SELFISH MINING

The organization of the blockchain is at risk of assaults of miners, who are mining for themselves. This selfish mining is considered a big challenge to this rising technology. Although a block is secured with proper validation, still is vulnerable to cheating with minimum hashing power. The miner holds with him the mined blocks without broadcasting on the blockchain network and making a private portion of the organization and communicated from that point satisfying certain necessities. This unduly work will make the legitimate miners worry about wasting time and resources with no use, because of the private network (chain) creation and mining procedure carryout by selfish miners, who are mining their blockchain without participants and generating more revenue. Chances are there as normal miners also get attracted to join the selfish pool and thereby exceeding 51% power quickly, blocks with less than 51% power are still at risk. Grounded on selfish mining, many other attacks are also possible only to show blockchain is not so sheltered. To resolve the selfish mining problem, a new technique is there or fair miners to decide which path to take. With the timestamp, honest miners would be able to select more new blocks that are again vulnerable to being fake or duplicate timestamp.

8.9 FUTURE ENHANCEMENT

Blockchain technology is still in the development phase. Decentralization, immutability, and transparency are the main pillars of blockchain. Much research is being led in this field to overcome challenges such as scalability, privacy leakage, selfish mining, and campaigning widespread for its global acceptance. Possible future directions concerning scalability are worked on how to optimize the storage in the block and redesigning the block-size with novel consensus algorithm-based solutions for scalability.

8.10 CONCLUSIONS

The central importance of blockchain innovation is to fabricate a robotized normal trust network that does not depend on outsiders and energizes the worth interconnection of the entire society. The benefits of blockchain incorporate decentralization, immutability, transparency, security, trust, and anonymity. The application areas have been discussed elaborately. Though blockchain excels in many areas, it faces certain challenges. These challenges never blocked the growth of blockchain technology.

REFERENCES

1. Regulation of Cryptocurrencies and Blockchain Technologies, National and International Perspectives, Rosario Girasa. Available at https://doi.org/10.1007/978-3-319-78509-7.
2. Peters G., Panayi E. Understanding Modern Banking Ledgers Through Blockchain Technologies: Future of Transaction Processing and Smart Contracts on the Internet of Money 2015. Available at SSRN: https://ssrn.com/abstract=2692487 or http://dx.doi.org/10.2139/ssrn.2692487.
3. https://www.coindesk.com/coindesk20.
4. Nakamoto S. Bitcoin: A Peer-to-Peer Electronic Cash System 2008. Available from: https://bitcoin.org/bitcoin.pdf.
5. Martínez G., Hernández-Álvarez L., Hernández Encinas L. Analysis of the Cryptographic Tools for Blockchain and Bitcoin, Published: 15 January 2020. Available at https://www.researchgate.net/publication/338648714_.
6. Begam S., Vimala J., Selvachandran G., Ngan T.T., Sharma R. Similarity Measure of Lattice Ordered Multi-Fuzzy Soft Sets Based on Set Theoretic Approach and Its Application in Decision Making. *Mathematics* 2020, 8, 1255.
7. Vo T., Sharma R., Kumar R., Son L.H., Pham B.T., Tien B.D., Priyadarshini I., Sarkar M., Le T. Crime Rate Detection Using Social Media of Different Crime Locations and Twitter Part-of-speech Tagger with Brown Clustering. 1 Jan. 2020, 4287–4299.
8. Nguyen P.T., Ha D.H., Avand M., Jaafari A., Nguyen H.D., Al-Ansari N., Van Phong T., Sharma R., Kumar R., Le H.V., Ho L.S., Prakash I., Pham B.T. Soft Computing Ensemble Models Based on Logistic Regression for Groundwater Potential Mapping. *Applied Science* 2020, 10, 2469.
9. Jha S. et al. Deep Learning Approach for Software Maintainability Metrics Prediction. *IEEE Access* 2019, 7, 61840–61855.
10. Sharma R., Kumar R., Sharma D.K., Son L.H., Priyadarshini I., Pham B.T., Bui D.T., Rai S. Inferring Air Pollution from Air Quality Index by Different Geographical Areas: Case Study in India. *Air Quality, Atmosphere, and Health* 2019, 12, 1347–1357.
11. Sharma R., Kumar R., Singh P.K., Raboaca M.S., Felseghi R.-A. A Systematic Study on the Analysis of the Emission of CO, CO_2 and HC for Four-Wheelers and Its Impact on the Sustainable Ecosystem. *Sustainability* 2020, 12, 6707.
12. Sharma S. et al. Global Forecasting Confirmed and Fatal Cases of COVID-19 Outbreak Using Autoregressive Integrated Moving Average Model. *Frontiers in Public Health* 2020. https://doi.org/10.3389/fpubh.2020.580327.
13. Malik P. et al. Industrial Internet of Things and its Applications in Industry 4.0: State-of the Art. *Computer Communication* 2021, 166, 125–139, Elsevier.
14. Analysis of Water Pollution using different Physico-Chemical Parameters: A Study of Yamuna River. *Frontiers in Environmental Science* 2020. https://doi.org/10.3389/fenvs.2020.581591.
15. Dansana D. et al. Using Susceptible-Exposed-Infectious-Recovered Model to Forecast Coronavirus Outbreak. *Computers, Materials & Continua* 2021, 67, no.2, 1595–1612.
16. Vo M.T., Vo A.H., Nguyen T., Sharma R., Le T. Dealing with the Class Imbalance Problem in the Detection of Fake Job Descriptions. *Computers, Materials & Continua* 2021, 68, no.1, 521–535.
17. Sachan S., Sharma R., Sehgal A. Energy Efficient Scheme for Better Connectivity in Sustainable Mobile Wireless Sensor Networks. *Sustainable Computing: Informatics and Systems* 2021, 30, 100504.
18. Ghanem S. et al. Lane Detection under Artificial Colored Light in Tunnels and on Highways: An IoT-Based Framework for Smart City Infrastructure. *Complex & Intelligent Systems* 2021. https://doi.org/10.1007/s40747-021-00381-2.

9 Mental Disorder Detection Using Machine Learning

Charu Chhabra
RDIAS

Sneha Chaudhary
JIMS

Ayasha Malik
Noida Institute of Engineering and technology

Bharat Bhushan
Sharda University

CONTENTS

9.1 INTRODUCTION

Machine learning (ML) is the discipline of computer science that deals in designing the intelligent machines. The technology lays emphasis in clutching on the promise to transform the healthcare and potential stumbling blocks. The deprecatory phase

DOI: 10.1201/9781003032328-9

of digital revolution which enormously deals with the coalescence of numerous technologies [1]. ML techniques are already functioning in almost every single domain and a lot of industries have fabricated these techniques into their working process. Healthcare is one such industry that has witnessed the sorcery of some astounding research works so far and promising the prodigious in coming years. Not just confined to a sector, ML has exploded with its capabilities and promising upshots in almost every possible sector. Healthcare which has now been transformed into smart healthcare has deployed the usage of ML has enabled an efficient, convenient and a personalized healthcare system [2]. The scope of healthcare has widely inflated to data associated with diseases, digital medical imaging, pattern recognition, cancer detection [3]. ML at present is being utilized to expedite early detection of diseases, facilitate better perception of disease progression, to ameliorate the treatment taking consideration of the medical condition and thereby unearth the possible treatments [4]. This article provides an overview of ML, Big Data Analytics (BDA), Artificial Intelligence (AI) and Deep Learning (DL) in medical healthcare also streamlining the mental disorders and psychiatric disorders followed by the remedies for the same. The terminology "ML" was conceived by the scientist John McCarthy depicted the term as the branch of computer science which mainly deals with the process of engineering intelligent machines [4]. It is appropriate to quote that ML is very much omnipresent in any and every application or industry [5]. At present ML is effectively utilized to keep a check on early disease detection and also empower and facilitate the better understanding of disease progression thereby upgrading and advancing medical treatment based on the analysis and requirements followed by future scope of applicability in numerous disciplines [6]. The most extensive tenacity of ML is its ability to rapidly analyze the pattern of tremendous data sets. The most important concept in medicine is the pattern recognition as based on the same early prediction and diagnosis could be executed by the clinical decision makers or physicians [7]. This is the era of Big Data (BD) and especially in the case of medicine and healthcare; there is the availability of plethora of data. BD is defined in general language; it can easily be termed out to be the concept which deals with managing the data which exceeds the processing activity and capacity of the conventional storage systems with analytics [8]. There is a principal component of the four Vs commonly termed as volume, veracity, velocity and variety, which is the major layout for workability in healthcare industry as the size of the medical data has rapidly grown and is increasing every single day [9]. There has been an observance which worked majorly on the capturing of hospital data wherein the case of 500 beds in a hospital had led to the evolution of data sets of somewhere more than 50 petabytes [10]. Therefore, BDA is impeccably being implemented in the medical healthcare sector as it not only controls and analyzes the data sets in general but as other V (veracity) it majorly deals with the close encounter of the real-time analysis and management based on the same. Not only this but also the effective working environment inculcating the concept of digital health records helps the proper management of the patient's database based on which it can conclude with a better patient doctor relationship. Digital records of patients contribute in analysis of the complete information either related to patient database or disease diagnosis related data, prescribed length of medication for instance and also costs of medication costs, testing related information either the case of Magnetic Resonance Imaging (MRI), Computerized Tomography-SCAN (CT-SCAN) or ultrasound, neuroimaging and radiology [11]. This provides an ease to the clinical experts and physicians to dive into the deep analysis and early prediction, diagnosis and prognosis of disease thereby yielding out a better efficient Clinical Decision Support System (CDSS). The splendid factor of ML in healthcare especially in relation with the Electronic Health Record (EHR) is that to enhance the capability of ML such that data sets can be analyzed in order to analyze the trends related to human behaviors which is an extremely challenging task for humans [12]. AI extensively contributes to the Decision Support Systems (DSS) as CDSS plays a vital role in healthcare informatics. Electronic records act as a complete knowledge base for DSS for hospitals and clinical experts as discussed above. In respect to the CDSS, it has been observed majorly that there has been a cutback in the expense of healthcare and also resulted in reduction of the length of patient stay in the hospital [13]. Also extending the feasibility of the CDSS the probability of revised admittance of patients has also greatly been reduced. There have been numerous contributions of AI and ML immensely proving out to be beneficial and efficient in healthcare, disease prediction, and further progressions with respect to the CDSS [14].

Owing to the widespread adoption of AI and ML, there have been a number of previously published surveys. For example, Farzi et al. [15] described Deep Belief Networks (DBNs) for prediction system in psychiatric disorder. Priya et al. [16] described the research of diagnosis of mental disorder based on the lifestyle and stress management of an individual. Sharma et al. [17] discussed the Unsupervised Learning (UL) and Supervised Learning (SL) methods inculcated for disease diagnosis. Hu et al. [18] described the Artificial Neural Network (ANN) which could be used for classification for early prediction. Tron et al. [19] discussed the K-means clustering in early prediction and analysis of the disorder in all possible causes and aspects. Xiong et al. [20] discussed about the Support Vector Machines (SVM) classifiers for filtering out the data and classifying the disease based on symptoms using ML algorithms [21].

Although ML algorithms have been extensively surveyed in numerous literature reviews which have been published but a few of the same have been identified and related [22]. There have been numerous publications in this discipline and several research gaps have been identified. Table 9.1 depicts the comparative study of the existing surveys in the research domain [23].

In order to fill the gaps observed, this paper presents the state-of-the-art ML technique to solve the security issues of mental disorder and psychiatric disorder.

The remaining section of the paper is organized as follows: Section 9.2 represents the background of the mental disorders and the possible causes of diagnosis. Section 9.3 highlights the research-oriented approach of ML techniques which have contributed to the disorder diagnosis and possible progression for treatment. Also, the ML techniques are discussed in the context of healthcare. Section 9.4 enlightens the common incurable bipolar disorder and detection codes. Section 9.5 describes a meta-analysis of depression. Section 9.6 depicts the diagnosis and prognosis of Parkinson's disease (PD). Section 9.7 contributes the research base on a meta-analysis of anxiety and ML techniques incorporated for the disease prediction. Section 9.8 showcases the cause and effects of schizophrenia which is a chronic mental disorder, where SVM is

TABLE 9.1

A Comparative Study and Summary of Existing Related Surveys

References	Publication Year	1	2	3	4	5	6	7	8
Tron et al. [19]	2016	-	Y	-	-	-	-	-	-
Sharma et al. [17]	2016	-	-	-	-	Y	-	-	-
Farzi et al. [15]	2017	Y	-	-	-	-	-	-	-
Hu et al. [18]	2020	Y	-	Y	-	-	Y	-	-
Xiong, Z et al. [20]	2020	-	-	-	Y	-	-	-	-
Priya et al. [16]	2020	-	-	-	-	-	Y	Y	-
Grande et al. [24]	2021	-	-	-	-	-	-	-	Y

1. Deep belief networks for prediction system.
2. K-means clustering analysis and clustering.
3. Convolutional neural networks for classification for early prediction.
4. Fuzzy clustering using statistical analysis.
5. Social media as a factor for developing disorder.
6. Disease prediction based on lifestyle.
7. Parkinson's and Alzheimer's chronic disease prediction and analysis.
8. Bipolar disorder detection.

used for prediction. Section 9.8 describes the schizophrenia and post-traumatic stress disorder. Finally, Section 9.9 concludes with conclusion [25].

9.2 A REVIEW OF MENTAL DISORDERS AND DISEASES

The application of ML has been anticipated in healthcare for quite a long now. ML has immensely contributed in various diagnosis and prognosis of various diseases. The techniques and studies assessed the prediction of the analytical fallouts and also in numerous cases UL and SVM have proved to be majorly used in diagnosis and prediction accordingly. The essence of mental illness remains an enigma till today. Yet, there has been the indulgence of psychiatrists and clinical experts who have laid immense emphasis and analysis in mapping to the alignment of the patterns of brain and behavioral characteristic of the patients suffering from mental disorder [26]. There has also been the indoctrination of the concepts of genomics and genetic analysis in ML for predicting the pattern of mental disorder. Either the case of BD or neurological disorders, or schizophrenia or suicidal, post-partum depression, Anorexia or Autism, it is observed that ML algorithms have been engaged in making predictions in psychiatry using genetics [15] as well. There have been numerous for framing the prediction models. These models have been built up using regression analysis either linear or logistic regression [16]. Not just confined to regression analysis, the concept of prediction in healthcare has been the most imperative part such that for such purpose the adoption of random forests, ANN, has also been exhibited [17]. ML is the subset of AI in general which mainly works out on mathematical models, statistical models in AI. Pattern matching and prediction are the most common characteristics of ML techniques as stated already.

It yields out the capability to excerpt the patterns and then mapping and analysis based on the observations in order to forecast the probable features and patterns. One such study reveals that ML algorithms have been successfully used for predicting the post-partum psychiatric disorder or depression at the time of neonatal exoneration [27]. It is nothing new in the fact that mother often is at an enormous peril and risk of evolving the post-partum depression followed by childbirth. They usually develop a mental disorder following childbirth and over a decade it has been observed that there has been a rapid increase in trend and analysis of the observance of maternal neonatal psychiatric disorder. ML has immensely contributed in predicting the early detection of such a condition as any severe psychiatric illness of the mother could lead to separation and estrangement of the mother and child units thereby affecting the cognitive and behavioral development of the infant. Not just confined till the neonatal deliver procedure, the process of prediction of the behavior and analysis has to be conducted post childbirth in order to distinguish the psychiatric admittance and also unearthing the possibility of depression, possibility, suicidal and other psychotic disorders [18]. Another condition of psychiatric disorder where ML algorithms have played a vital role in prediction and diagnosis of early prediction of disease model has been Autism Spectrum Disorder (ASD). The ML techniques widely used in the brain disorder and prediction of severity of the medical status have been for instance naïve Bayes', random forests, SL and UL, SVM, k-Nearest Neighbors (K-NN), ANN and Data Learning (DL) models and many to the count. The application of ML to neural disorders gigantically leans onto the SL algorithms which basically deal with training and teaching the machine thereby using the data which is labeled and somehow the data are tagged indirectly with the correct output data [28]. There has been an immense emphasis on classification, prediction and regression analysis in this context. Mental disorder diagnosis can be of two types. First is the prediction of the disease and the other is forecasting the relative outputs based on the pattern mapping using various ML algorithms [29].

9.3 MACHINE LEARNING FOR DISORDER DIAGNOSIS AND PROGNOSIS

ML is the subset of AI which in core deals with the concept of constructing intelligent machines. It is evident from the key terminology that ML mainly confines out to the development of mimic of human cognitive behavior of machines. The disciplinal learning in ML is classified into wide categories in general, namely, SL, UL, Reinforcement Learning (RL) and Semi-Supervised Learning (SSL) [30].

9.3.1 SUPERVISED LEARNING APPROACH

The SL approach is the category of ML techniques which deal with the concept of a supervisor guiding with the possible inputs and labels to yield out appropriate outputs. In general, SL is the type which works in such a way that the algorithms learn from the object- or case-related data, and there are tags or data associations in such a way that the prediction can easily be exhibited based on the tagged data. It may also relate to be one of the types which works under the supervision of a teacher guiding with principles to yield out appropriate outputs [31].

Neehal et al. [32] discussed the workability of SL techniques in the diagnosis of PD. The research stated the methodologies which used functional Magnetic Resonance Imaging (fMRI) data for conducting the research. Also, SVM classifier was used for the classification and prediction of PD. The major contribution of the research laid onto this process which turned out to be a well-structured model for predicting the early stages of PD. It may help the doctors for diagnosis of the disease at its early stages and the patients should receive better treatment. Moradi et al. [33] proposed a research focused on the use of SSL for predicting Alzheimer's disease (AD) conversion in Mild Cognitive Impairment (MCI) patients based on MRI. SSL methods differ from standard SL methods in that they make use of unlabeled data – in this case data from MCI subjects whose final diagnosis is not yet known. We compare two widely used semi-supervised methods, Low Density Separation (LDS) and Semi-Supervised Discriminant Analysis (SDA), to the corresponding supervised methods using real and synthetic MRI data of MCI subjects. With simulated data, using SSL instead of SL led to higher classification performance in certain cases; however, the applicability of semi-supervised methods depended strongly on the data distributions. With real MRI data, the SSL methods achieved significantly better classification performances over SL methods. Moreover, even using a small number of unlabeled samples improved the AD conversion predictions [34]. Figure 9.1 shows the main categories of SL.

9.3.2 UNSUPERVISED LEARNING APPROACH

In this ML approach, there is no requirement of any supervisor in the model. That means it can be stated out as a model which no labels are given in the algorithmic model and hence the machine predicts based on the available and unknown data sets with analytics [35]. Figure 9.2 states the different types or categories of UL algorithms in general.

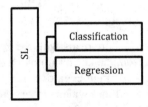

FIGURE 9.1   Categories of SL.

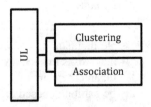

FIGURE 9.2   Classification of UL techniques.

9.3.3 REINFORCEMENT LEARNING APPROACH

The category of ML techniques which focuses solely on the rewards-based actions is stated as RL. In the same the intelligent agents take necessary actions in an environment in order to increase or maximize the rewards.

The other learning, i.e., the UL as the name suggests is the one which algorithm works without any supervision or guidance of a teacher or may be the one works on its experiential learning. Technically this type of ML deals with no data association and enables the algorithms to learn through the patterns on its own and make the prediction accordingly. The third type of ML algorithms is the RL which mainly focuses on reward-based learning in general [36]. In simple language, it can be termed as the hit and trial methodology, wherein the machine algorithm that learns based on the feedback, yes or no or may be good or bad, has been rewarded and learns accordingly. In the last category of SSL, it is basically one of the types which forms a hybrid orientation of both SL and UL in such a way that herein it uses a consolidation of cramped amount of labeled data and considerable amount of unlabeled data which works immensely in convoying the models. In today's era of social media, there has been an observance of bipolar disorder which has been encountered by the virtue of blooming social media use, stress levels, pressure on the minds of the people, and competition [37]. It is a study which reveals that this disease is rather mis-investigated as Major Depressive Disorder (MDD). The bipolar disorder in many cases has not been mentioned or reported thereby causing the person to remain influenced by the disease during the lifetime. There have been rigorous ML models involved in detecting the same at the primary level by either diagnosing the symptoms of presence of bipolar disorder. It is now relatively easy to state that machine learning has been a technology by the virtue of which the problems of clinical psychology and psychiatry can be resolved appropriately as it categorizes itself into steps such as diagnosis, prognosis, treatment prediction, potential biomarkers [38].

9.4 POSSIBLE CAUSES OF BIPOLAR DISORDER AND DETECTION CODES

The bipolar disorder is a very protracted disorder which is generally a mood and behavior changing disorder which is statistically affecting somewhere around 2% of the total adult population [39]. There is no such specific reason but majorly it is estimated and observed that it is accounted more on the high socio-economic burden majorly. In general, it is termed as mood disorder an umbrella term but the term itself comprises of many different types including depression which we shall be discussing in the next segment and bipolar disorder for exhaustive study. The mental health professionals match the pattern and analyze the symptoms of the affected person either an adult, a child or adolescent. MDD or bipolar disorder is the major psychiatric disorder which acts extremely devastating for the patient and family members. The devastating part of the disorder is the fact that it affects the patient's social and psychological functioning and the aspect of life in totality either we take the psychosocial or socio-economic condition [40].

There have been clinical trials for the patients in order to cure the bipolar disorder but the randomized clinical trials as per the research have contributed immensely in recommending effective medication, prescription and prognosis overall for bipolar disorder. The treatment methods such as Lamotrigine and Quetiapine have been effective by using statistical methods thereby providing results effectively so far. Till now, the summarization and analytics by the clinical experts or the mental health experts is being executed based on the symptoms base approach which is not as effective in many real time scenarios. This could lead to the disparity and divergence in the diagnostic approach. To conclude more emphasis shall be laid onto the early diagnosis with clinical trials and analysis along with the requirement of clinical expertise and mental health professionals with the specialization will be highly needed. Passing on to this requirement now how does ML can eventually contribute in early diagnosis and progressions, it is very appropriate to state that ML is the field of AI which is when applied in big data analytics utilizes analytical mathematical models and statistical models in order to supervise the learning models and machines [41]. The most incredible part of ML is that it ultimately encounters and unearth the complex patterns and behaviors using asymptotic analysis and further lays out a predictive model or forecast and/or predictions based on even the new data or alien data or unseen data. Lish et al. [42] explained the effectiveness of the current methodologies adopted for diagnosing the bipolar disorder as they were many challenges in the current diagnostic treatments methods which needed to be transformed. Sackett et al. [43] showcased another research evaluation wherein evidence-based medicine worked as a base to understand the possible aspects of risks, superlative treatments and also prognosis of bipolar disorder by using the statistical methods. Huys et al. [44] presented in the light of the above findings the techniques which aim at the pertinent psychiatric care as ML has been gaining grounds in the psychiatric disorder researches. Salomon et al. [45] presented the details relatively generated by Healthy Life Expectancy (HLE) which summarized the mortality and non-fatal outcomes in a single measure of average population health. It has been used to compare health between countries or to measure changes over time. The comparisons incurred the inferences that enabled the system inform and aware about policy questions that depend on how morbidity changes as mortality decreases. With an extension to the above observation, Gao et al. [46] concluded that the patients who are diagnosed with the mood disorder have to sometimes in many cases endure the wrong drug trial or possibly multiple trials before receiving a final diagnosis. In such a case, the diagnostic statistical manual classification is relatively unclear and thereby creates a confusion in predicting the exact type of mood disorder or the reason of the mood disorder for effective diagnosis and prognosis.

9.5 A META-ANALYSIS OF DEPRESSION

Depression is a highly extensive and ubiquitous psychiatric disorder which majorly impacts the life of the patient and also for the family it significantly affects by laying socio-economic burden. There are strategies in which the image processing in neuroimaging provided measurements of brain functioning and structure set up. This helps in understanding the basic descriptive biomarkers. Chen et al. [47] concluded this theory effectively wherein the discussion was effectively laid on the adaptation of biomarkers for effective diagnosis. In the recent research findings, the activities followed by which the classification of patients' disorders has been executed has been set up in categories and based on the same inferences have been drawn out using ML techniques majorly naive Baye's classifiers and SVM in order to model the activities and based on that to predict the mood outcome.

9.6 DIAGNOSIS AND PROGNOSIS OF PARKINSON'S DISEASE

PD is another major disorder which is in general a neurodegenerative disorder, and on records, it is considered to be the second most prevalent in the world. Muthusamy et al. [48] highlighted the role of technological tools in the PD field for patient self-reports, diaries, PD assessments and unified PD rating scale. There has no defined treatment or prognosis for PD yet which can effectively contribute in reduction in disease progression. Earlier in the previous research contributions, the classification and detection based on the demonstration on the basis of motor symptoms were exhibited. These motor symptoms included stiffness of the muscles, tremors, slow movements and also basic balance movement. Prior to this, the major emphasis was laid out in predicting and classifying the symptoms and clustering them also. For working out efficiently in this domain, computational intelligence techniques were employed. Also, ANN was effectively used in the making to capture the neuroimaging-based brain mapping to visualize how the neurons were operating in the brain to process the data retrieved from the senses, basic functioning of the neurons. Zhang et al. [49] summarized some of the most important developments in neural network classification research. In this research contribution, it was elaborated considering classification to be the most important and frequently confronted decision-making effort of human activity. For the same, several neural network models were discussed in for the research problems and also explained several theoretical ad empirical issues of neural networks were elaborated. Olivares et al. [50] proposed a model in which they used the PD audio data set taken from ML repository. ELM is stated as a feed forward neural architectural setup in which the input weights and hidden layers need not be computed iteratively [38].

9.7 A META-ANALYSIS OF ANXIETY

Anxiety is a disorder which can be measured in many ways which includes self-evaluated record, considered parameters, and behavioral measures. In today's era, anxiety has been the most common mental disorder owing to our lifestyle and stress management. Several researches in the past have been exhibited with the reason and early prediction and further progression and management of the disorder. Rohini et al. [51] proposed the research metrics which were oriented in predictions of anxiety.

Ml algorithms were used to predict the stress and depression. It stated the fact that these disorders were predicted based on different levels of intensity of severity using ML algorithms. The algorithms contributed for the early prediction were namely Decision Trees (DT), KNN, naïve Baye's and SVM. Also, as an observation in the case scenario of anxiety, it can also be case to predict the anxiety via a model of the ratio of population thereby depicting the measurement of the levels of depression of the people by wither using or mapping with their online blogs or actions or tweets possibly [52].

Hinton et al. [47] proposed a research contribution evaluating the age factors and possible reasons for developing the anxiety. The research stated the possible methodologies which could enable an earlier diagnosis of anxiety in children specially where in the situation usually arises in the form when a child probably cannot sit calmly and unable to focus and control its behavior. The results showed the superiority of the proposed method rather than other systems. McGinnis et al. [53] proposed an approach for diagnosing anxiety and depression in young children. At present the diagnosis requires hours of framed and processed clinical interviews and standard set of questionnaires passes on over days or weeks. Also stated the use of a 90-second fear induction task during which time participant motion is monitoring using a commercially available wearable sensor. Chang et al. [54] surveyed the effects of Social Anxiety Disorder (SAD), one of the most important and common mental disorders, affecting a large proportion of individuals around the world. The prediction of the disorder in patients was carried by the ML algorithms and also data analytics from the social media activities to extract multiple effective features for early diagnosis systems was carried out.

Basically, they are statistical models which depict the relationship between the input units and output unit. Image processing and radiology forms a major role in pattern recognition. It incorporates major contribution in speech recognition, image processing, machine translation and many to the counting. Simms et al. [55] proposed a model in which they followed in the tradition and major our attention was given in on cognitive distortion, a precursor and symptom of disruptive psychological disorders such as anxiety, anorexia and depression. Moreover, a number of personal blogs were collected and labeled based on whether they exhibited distorted thought patterns. The overall observations showcased that it is possible to detect cognitive distortions automatically from personal blogs with relatively good accuracy (73.0%) and false negative rate (30.4%).

9.8   SCHIZOPHRENIA AND POST-TRAUMATIC STRESS DISORDER

Schizophrenia is an acute chronic mental disorder. When active, the disease can cause symptoms like delusions and hallucinations. Though as such there is no cure for the disease, there are findings for better innovative solutions and safer treatments. There are also theories wherein experts also are unraveling the causes of the disease by studying genetics, conducting behavioral research, and using advanced imaging to look at the brain's structure and function. These approaches hold the promise of new and more effective therapies.

Almutairi et al. [56] proposed a research whose purpose was to develop a predictive system to preemptively diagnose schizophrenia disease using computational intelligence-based techniques. The research showcases all the possibilities of getting the disease at an early stage, which will improve the health state of the patients. The exact exhibition was done using machine learning techniques. Swati et al. [57] discussed the technology evolution in schizophrenia detection using ML-based classification algorithms. Similarly, Hubbard et al. [58] discussed the role of ML-based neuroimaging techniques dedicated towards the brain mapping disorders. The work stressed on the issues related to over-fitting and high dimensional input.

9.8.1 Support Vector Machine for Prediction

SVMs are basically a set of SL methods which are in general used for classification, regression and Outlier Detection (OD). There are many advantages of SVM for prediction analysis in disease diagnosis and prognosis. The SVM has tremendously been used as an ultimate classifier in disease prediction and diagnosis. Some of these are breast cancer, psychiatric disorder, and diabetes. They are effective in very high domain and dimensions of data relatively. Also, in some cases, the dimensions are far more than the sample stores of the data.

Pflueger et al. [59] discussed the capability of Convolutional Neural Network (CNN) in order to identify the patients diagnosed with Schizophrenia using 3D brain MR images followed by finding an alternative for imaging-based diagnosis and prognosis of psychiatric disorder. Katarya et al. [60] used a light-depth camera and dedicated software to automatically measure the facial muscular activity of Schizophrenia patients and healthy individuals during a short interview. This record mapping was entirely based on K-means clustering analysis; facial activity was characterized in terms of typicality, richness and distribution of seven facial-clusters [61].

9.9 CONCLUSION

Mental health maps the social well-being, emotional and the obviously psychological well-being of an entity or an individual. It indicates the ability of an individual to handle stress, make decisions, handling situations and overall, how does an individual feel. It is appropriate to consider that mental health concern like physical health is necessary to be emphasized at every phase in the life being childhood or later stages. The mental health state of an individual can often affect the way one perceives, mood and also entirely the behavior. There have been tremendous researches for the applicability of ML algorithms and techniques and DL techniques. Recently, machine learning has been widely used in the areas of neurological disorders, radiology, exploring the image diagnosis and prognosis in disease pattern matching, AD/MCI, tumor, schizophrenia, PD, cardiac stroke and traumatic brain injuries, post-partum depression. There has been a significant increase in usage of ML techniques in keeping the track of brain disorders in neuroimaging data. We have discussed the numerous factors responsible for mental health problems, thereby leading to mental illness.

REFERENCES

1. Pang, Z., Yuan, H., Zhang, Y., Packirisamy, M. (2018). Guest editorial health engineering driven by the industry 4.0 for aging society. *IEEE Journal of Biomedical and Health Informatics*, 22(6), 1709–1710. doi:10.1109/jbhi.2018.2874081.
2. Khedkar, S., Subramanian, V., Shinde, G., Gandhi, P. (2019). Explainable AI in healthcare. *SSRN Electronic Journal*. doi:10.2139/ssrn.336768.

3. Arisi, I., Mecocci, P., Bruno, G., Canevelli, M., Tsolaki, M., Pelteki, N., Cattaneo, A. (2018). Mining clinical and laboratory data of neurodegenerative diseases by machine learning: Transcriptomic biomarkers. *2018 IEEE International Conference on Bioinformatics and Biomedicine (BIBM)*. doi:10.1109/bibm.2018.8621072.

4. McCarthy, J. (1989). Artificial intelligence, logic and formalizing common sense. *Philosophical Logic and Artificial Intelligence*, 161–190. doi:10.1007/978-94-009-2448-2_6.

5. Beam, A.L., Kohane, I.S. (2016). Translating artificial intelligence into clinical care. *JAMA*, 316(22), 2368–2369. https://doi.org/10.1001/ jama.2016.17217.

6. Bishnoi, L., Narayan Singh, S. (2018). Artificial intelligence techniques used in medical sciences: A review. *2018 8th International Conference on Cloud Computing, Data Science & Engineering (Confluence)*. doi:10.1109/confluence.2018.8442729.

7. Fogel, A.L., Kvedar, J.C. (2018). Artificial intelligence powers digital medicine. *Npj Digit Medicine*, 1(1), 3–6. https://doi.org/10.1038/ s41746-017-0012-2.

8. Dumbill, E. (2013). Making sense of big data. *Big Data*, 1(1), 1–2. doi:10.1089/ big.2012.1503.

9. Begam, S., Vimala, J., Selvachandran, G., Ngan, T.T., Sharma, R. (2020). Similarity measure of lattice ordered multi-fuzzy soft sets based on set theoretic approach and its application in decision making. *Mathematics*, 8, 1255.

10. Ghesu, F.C., Krubasik, E., Georgescu, B., Singh, V., Zheng, Y., Hornegger, J., Comaniciu, D. (2016). Marginal space deep learning: Efficient architecture for volumetric image parsing. *IEEE Transactions on Medical Imaging*, 35(5), 1217–1228. doi:10.1109/tmi.2016.2538802.

11. Vo, T., Sharma, R., Kumar, R., Son, L.H., Pham, B.T., Tien, B.D., Priyadarshini, I., Sarkar, M., Le, T. (2020). Crime rate detection using social media of different crime locations and twitter part-of-speech tagger with brown clustering, 4287–4299.

12. Wang, Y., Kung, L.A., Byrd, T.A. (2016). Big data analytics: Understanding its capabilities and potential benefits for healthcare organizations. *Technological Forecasting and Social Change*, 126, 3–13. https://doi.org/ 10.1016/j.techfore.2015.12.019.

13. Jadhav, R., Chellwani, V., Deshmukh, S., Sachdev, H. (2019). Mental disorder detection: Bipolar disorder scrutinization using machine learning. *2019 9th International Conference on Cloud Computing, Data Science & Engineering (Confluence)*. doi:10.1109/confluence.2019.8776913.

14. Nguyen, P.T., Ha, D.H., Avand, M., Jaafari, A., Nguyen, H.D., Al-Ansari, N., Van Phong, T., Sharma, R., Kumar, R., Le, H.V., Ho, L.S., Prakash, I., Pham, B.T. (2020). Soft computing ensemble models based on logistic regression for groundwater potential mapping. *Applied Science*, 10, 2469.

15. Farzi, S., Kianian, S., Rastkhadive, I. (2017). Diagnosis of attention deficit hyperactivity disorder using deep belief network based on greedy approach. *2017 5th International Symposium on Computational and Business Intelligence (ISCBI)*. doi:10.1109/ iscbi.2017.8053552.

16. Priya, A., Garg, S., Tigga, N.P. (2020). Predicting anxiety, depression and stress in modern life using machine learning algorithms. *Procedia Computer Science*, 167, 1258–1267. doi:10.1016/j.procs.2020.03.442.

17. Sharma, V., Prakash, N.R., Kalra, P. (2016). EDA wavelet features as social anxiety Disorder (SAD) estimator in adolescent females. *2016 International Conference on Advances in Computing, Communications and Informatics (ICACCI)*. doi:10.1109/ icacci.2016.7732317.

18. Hu, M., Sim, K., Zhou, J. H., Jiang, X., Guan, C. (2020). Brain MRI-based 3D convolutional neural networks for classification of schizophrenia and Controls. *2020 42nd Annual International Conference of the IEEE Engineering in Medicine & Biology Society (EMBC)*. doi:10.1109/embc44109.2020.9176610.

19. Tron, T., Peled, A., Grinsphoon, A., Weinshall, D. (2016). Facial expressions and flat affect in schizophrenia, automatic analysis from depth camera data. *2016 IEEE-EMBS International Conference on Biomedical and Health Informatics (BHI)*. doi:10.1109/bhi.2016.7455874.

20. Xiong, Z., Zhang, X., Zhang, M., Cao, B. (2020). Predicting features of human mental disorders through methylation profile and machine learning models. *2020 2nd International Conference on Machine Learning, Big Data and Business Intelligence (MLBDBI)*. doi:10.1109/mlbdbi51377.2020.00019.

21. Jha, S. et al. (2019). Deep learning approach for software maintainability metrics prediction. *IEEE Access*, 7, 61840–61855.

22. American psychiatric Association diagnostic and Statistical Manual of mental Disorders (DSM-IV). (n.d.). SpringerReference. doi:10.1007/springerreference_179660.

23. Sharma, R., Kumar, R., Sharma, D.K., Son, L.H., Priyadarshini, I., Pham, B.T., Bui, D.T., Rai, S. (2019). Inferring air pollution from air quality index by different geographical areas: Case study in India. *Air Quality, Atmosphere, and Health*, 12, 1347–1357.

24. Grande, I., Berk, M., Birmaher, B., Vieta, E. (2016). Bipolar disorder. *The Lancet*, 387(10027), 1561–1572. doi:10.1016/s0140-6736(15)00241-x.

25. Sharma, R., Kumar, R., Singh, P.K., Raboaca, M.S., Felseghi, R.-A. (2020). A systematic study on the analysis of the emission of CO, CO_2 and HC for four-wheelers and its impact on the sustainable ecosystem. *Sustainability*, 12, 6707.

26. Sharma, S. et al. (2020). Global forecasting confirmed and fatal cases of COVID-19 outbreak using autoregressive integrated moving average model. *Frontiers in Public Health*. https://doi.org/10.3389/fpubh.2020.580327.

27. Malik, P. et al. (2021). Industrial internet of things and its applications in industry 4.0: State-of the art. *Computer Communication*, 166, 125–139, Elsevier.

28. Analysis of water pollution using different physico-chemical parameters: A study of Yamuna river. *Frontiers in Environmental Science*. https://doi.org/10.3389/fenvs.2020.581591.

29. Geddes, J.R., Calabrese, J.R., Goodwin, G.M. (2009). Lamotrigine for treatment of bipolar depression: Independent meta-analysis and meta-regression of individual patient data from five randomized trials. *British Journal of Psychiatry*, 194(1), 4–9. https://doi.org/10.1192/bjp. bp.107.048504.

30. Young, A.H., McElroy, S.L., Olausson, B., Paulsson, B. (2014). A randomised, placebo-controlled 52-week trial of continued quetiapine treatment in recently depressed patients with bipolar I and bipolar II disorder. *World Journal of Biological Psychiatry*, 15(2), 96–112. https://doi. org/10.3109/15622975.2012.665177.

31. Wang, J., Chen, Q., Jiang, Z., Li, X., Kuang, H., Chen, T., Chen, H. (2020). Withdrawn: Epidemiology and clinical analysis of the breakout of dengue fever in Zhangshu City, Jiangxi province in 2019. *Médecine Et Maladies Infectieuses*. doi:10.1016/j.medmal.2020.09.017.

32. Neehal, A.H., Azam, M.N., Islam, M.S., Hossain, M.I., Parvez, M.Z. (2020). Prediction of Parkinson's disease by analyzing fMRI data and using supervised learning. *2020 IEEE Region 10 Symposium (TENSYMP)*. doi:10.1109/tensymp50017.2020.9230918.

33. Moradi, E., Tohka, J., Gaser, C. (2014). Semi-supervised learning IN mci-to-ad conversion prediction — when is unlabeled data useful? *2014 International Workshop on Pattern Recognition in Neuroimaging*. doi:10.1109/prni.2014.6858535.

34. Hruschka, E., De Castro, L., Campello, R. (n.d.). Evolutionary algorithms for clustering gene-expression data. *Fourth IEEE International Conference on Data Mining (ICDM'04)*. doi:10.1109/icdm.2004.10073.

35. Ding, J., Condon, A., Shah, S.P. (2017). Interpretable dimensionality reduction of single cell transcriptome data with deep generative models. *Nature Communications*. doi:10.1101/178624.

36. Dansana, D. et al. (2021). Using susceptible-exposed-infectious-recovered model to forecast coronavirus outbreak. *Computers, Materials & Continua*, 67(2), 1595–1612.
37. Wong, K.Y., Chung, F. (2019). Visualizing time series data with temporal matching based t-sne. *2019 International Joint Conference on Neural Networks (IJCNN)*. doi:10.1109/ijcnn.2019.8851847.
38. Vo, M.T., Vo, A.H., Nguyen, T., Sharma, R., Le, T. (2021). Dealing with the class imbalance problem in the detection of fake job descriptions. *Computers, Materials & Continua*, 68(1), 521–535.
39. Singh, G., Vadera, M., Samavedham, L., Lim, E.C. (2016). Machine learning-based framework for multi-class diagnosis of neurodegenerative diseases: A study on parkinson's disease. *IFAC-PapersOnLine*, 49(7), 990–995. doi:10.1016/j.ifacol.2016.07.331.
40. Sachan, S., Sharma, R., Sehgal, A. (2021). Energy efficient scheme for better connectivity in sustainable mobile wireless sensor networks. *Sustainable Computing: Informatics and Systems*, 30, 100504.
41. Kessing, L.V., Andersen, P.K., Vinberg, M. (2017). Risk of recurrence after a single manic or mixed episode – A systematic review and meta-analysis. *Bipolar Disorders*, 20(1), 9–17. doi:10.1111/bdi.12593.
42. Lish, J.D., Dime-Meenan, S., Whybrow, P.C., Price, R., Hirschfeld, R.M. (1994). The national depressive and Manic-depressive association (DMDA) survey of bipolar members. *Journal of Affective Disorders*, 31(4), 281–294. doi:10.1016/0165-0327(94)90104-x.
43. Sackett, D.L., Rosenberg, W.M., Gray, J.A., Haynes, R.B., Richardson, W.S. (1996). Evidence based medicine: What it is and what it isn't. *BMJ*, 312(7023), 71–72. doi:10.1136/bmj.312.7023.71.
44. Huys, Q.J., Maia, T.V., Frank, M.J. (2016). Computational psychiatry as a bridge from neuroscience to clinical applications. *Nature Neuroscience*, 19(3), 404–413. doi:10.1038/nn.4238.
45. Salomon, J.A., Wang, H., Freeman, M.K., Vos, T., Flaxman, A.D., Lopez, A.D., Murray, C.J. (2012). Healthy life expectancy for 187 COUNTRIES, 1990–2010: A systematic analysis for the Global Burden Disease study 2010. *The Lancet*, 380(9859), 2144–2162. doi:10.1016/s0140-6736(12)61690-0.
46. Gao, S., Osuch, E.A., Wammes, M., Theberge, J., Jiang, T., Calhoun, V.D., Sui, J. (2017). Discriminating bipolar disorder from major depression based on kernel SVM using FUNCTIONAL independent components. *2017 IEEE 27th International Workshop on Machine Learning for Signal Processing (MLSP)*. doi:10.1109/mlsp.2017.8168110.
47. He, Y., Chen, Z.J., Evans, A.C. (2007). Small-world anatomical networks in the human brain revealed by cortical thickness from MRI. *Cerebral Cortex*, 17(10), 2407–2419. doi:10.1093/cercor/bhl149.
48. Oung, Q., Muthusamy, H., Lee, H., Basah, S., Yaacob, S., Sarillee, M., & Lee, C. (2015). Technologies for assessment of motor disorders in Parkinson's disease: A review. *Sensors*, 15(9), 21710–21745. doi:10.3390/s150921710.
49. Zhang, G. (2000). Neural networks for classification: A survey. *IEEE Transactions on Systems, Man and Cybernetics, Part C (Applications and Reviews)*, 30(4), 451–462. doi:10.1109/5326.897072.
50. Olivares, R., Munoz, R., Soto, R., Crawford, B., Cárdenas, D., Ponce, A., Taramasco, C. (2020). An optimized brain-based algorithm for classifying Parkinson's disease. *Applied Sciences*, 10(5), 1827. doi:10.3390/app10051827
51. Rohani, D.A., Springer, A., Hollis, V., Bardram, J.E., Whittaker, S. (2020). Recommending activities for mental health and well-being: Insights from two user studies. *IEEE Transactions on Emerging Topics in Computing*, 1. doi:10.1109/tetc.2020.2972007
52. Ghanem, S. et al. (2021). Lane detection under artificial colored light in tunnels and on highways: An IoT-based framework for smart city infrastructure. *Complex & Intelligent Systems*. https://doi.org/10.1007/s40747-021-00381-2.

53. McGinnis, R.S., McGinnis, E.W., Hruschak, J., Lopez-Duran, N.L., Fitzgerald, K., Rosenblum, K.L., Muzik, M. (2018). Rapid anxiety and depression diagnosis in young Children enabled by wearable sensors and machine learning. *2018 40th Annual International Conference of the IEEE Engineering in Medicine and Biology Society (EMBC)*. doi:10.1109/embc.2018.8513327.
54. Chang, M., Tseng, C. (2020). Detecting social anxiety with online social network data. *2020 21st IEEE International Conference on Mobile Data Management (MDM)*. doi:10.1109/mdm48529.2020.00073.
55. Simms, T., Ramstedt, C., Rich, M., Richards, M., Martinez, T., Giraud-Carrier, C. (2017). Detecting cognitive distortions through machine learning text analytics. *2017 IEEE International Conference on Healthcare Informatics (ICHI)*. doi:10.1109/ichi.2017.39.
56. Almutairi, M.M., Alhamad, N., Alyami, A., Alshobbar, Z., Alfayez, H., Al-Akkas, N., … Olatunji, S.O. (2019). Preemptive diagnosis of schizophrenia disease using computational intelligence techniques. *2019 2nd International Conference on Computer Applications & Information Security (ICCAIS)*. doi:10.1109/cais.2019.8769513.
57. Swati, N., Indiramma, M. (2020). Machine learning systems for detecting schizophrenia. *2020 Fourth International Conference on I-SMAC (IoT in Social, Mobile, Analytics and Cloud) (I-SMAC)*. doi:10.1109/i-smac49090.2020.9243597.
58. Hubbard, A., Van der Laan, M. (2020). Developing methods to predict health outcomes in trauma patients. doi:10.25302/1.2020.me.130602735.
59. Pflueger, M.O., Franke, I., Graf, M., Hachtel, H. (2015). Predicting general criminal recidivism in mentally disordered offenders using a random forest approach. *BMC Psychiatry*, 15(1). doi:10.1186/s12888-015-0447-4.
60. Katarya, R., Maan, S. (2020). Predicting mental health disorders using machine learning for employees in technical and non-technical companies. *2020 IEEE International Conference on Advances and Developments in Electrical and Electronics Engineering (ICADEE)*. doi:10.1109/icadee51157.2020.9368923.
61. De Pierrefeu, A., Lofstedt, T., Laidi, C., Hadj-Selem, F., Leboyer, M., Ciuciu, P., Duchesnay, (2018). Interpretable and stable prediction of schizophrenia on a large multisite dataset using machine learning with structured sparsity. *2018 International Workshop on Pattern Recognition in Neuroimaging (PRNI)*. doi:10.1109/prni.2018.8423946.

10 Blockchain Technology for Industry 4.0 Applications

Issues, Challenges and Future Research Directions

Himanshu, Rishu Rana & Nikhil Sharma
HMR Institute of Technology & Management, Delhi, India

Ila Kaushik
Krishna Institute of Engineering &
Technology, Ghaziabad, India

Bharat Bhushan
School of Engineering and Technology,
Sharda University, Greater Noida, India

CONTENTS

DOI: 10.1201/9781003032328-10

10.1 INTRODUCTION

In today's era two popular smart contracts, i.e., Bitcoin and Ethereum are well recognized for the execution of blockchains. When there is continual growth in a list that contains unchangeable data records, this leads to the formation of Latter [1]. Ledger (public) is used to store information of each transaction and to record data within the blockchain and that information is not stored in an abrupt manner. It is started in a dispensed ledger which is shared among all participating nodes. Blockchain is highly used for transaction management between different parties via peer to peer (P2P) network [2]. In other words, it provides unyielding, faithful, and interchangeable data, consorts, and performs transactions. As blockchain is open-source and no central community has the right to control this, before accepting a block for involvement into the ledger, the existing nodes have to gain an agreement by running a prebuilt agreement algorithm used in the protocol of blockchain and they are as follows: PoS (Proof-of-Stake), DPoS (Delighted-Proof-of-Stake), PBFT (Practical-Byzantine-Fault-Tolerance), PoW (Proof-of-Work), PoA (Proof-of-Authority/Proof-of-Authenticity) [3], PoC (Proof-of-Concept), PoE (Proof-of-Existence), PoO (Proof-of-Object), PoG (Proof-of-Graph) [4] and PoD (Proof-of-Data) [5]. There are several fundamental applications of blockchain which are very helpful in today's world, with trust one could securely store their financial data on the blockchain, many researchers have been practicing to enlarge the applications of blockchain beyond payments to other sectors like healthcare, education, and supply chain manufacturing [6].

Blockchain 1.0 is generally associated with payment and cryptocurrency. An example of this is bitcoin. Blockchain acts as a base for bitcoin, with the help of the blockchain the central authorities are not required to operate bitcoin and other cryptocurrencies. Along with decreasing the risk, it also helps to abolish transaction and processing fees [7].

Blockchain 2.0 is linked with programmed digital finance with the help of smart arrangements. An example of this is bitcoin, which is the updated version of bitcoin, which offers much transaction speed than bitcoin. Gradually, the demand for bitcoin is increasing due to the increase in its usage rates [8].

Blockchain 3.0 is a highly trending one and focuses on the requirement of digital society. This is the upgraded version of blockchain 2.0. In order to solve existing scalability problems, blockchain 3.0 is assembled which provides more speedy transactions and cost-effectiveness. Examples of these technologies are smart cities and Industry 4.0. Industry 4.0 is the networking of systems and processes for industries directly linked with IoT (Internet of Things). Basically, Industry 4.0 is the fourth industrial revolution. Industry 4.0 was firstly defined by the German government in 2011 [9], which is now highly accepted by different sectors. For example, Industrial Internet of Things (IIoT) is similar to IoT or we can say that IIoT is a subdivision of IoT. There are different methods or attempts to use blockchain in different sectors like the IIoT sector and it is used to manage the data collection. We have also gone through different research papers and assessments of blockchain applications in industry, smart cities and IoT. For example, authors of assessment discussed different advantages and disadvantages, usage of Blockchain in IoT. A limited number of estimations and assessments focused on implementation of blockchain in specific industries or implementation based on edge computing and so on [10,11]. In sectors like financial services, insurance, manufacturing, healthcare, government, transportation, etc., the IoT has grown. In these sectors, the Industrial IoT (IIoT) focuses on the use of IoT, which joins emerging technologies such as artificial intelligence, smart sensors, robots, machine-to-machine (M2M) and much more into traditional industrial procedures [12]. Modern industries regularly look for the factors that enhance the capabilities, and these factors are flexible, system stability and cost effectiveness [13]. So, cloud computing and IoT are combined with each other so that it fulfils the industry need. It is expected that due to increasing IIoT, there is also growth in the economic sector and creation of next-generation smart systems [14]. Due to interconnecting facilities, the IIoT is dramatically changing the way of work of industries and creating fresh content for the business [15]. While using IoT devices, a huge amount of security-sensitive and confidential material is generated. According to CISCO, 50 billion devices would grab the Internet by 2021, and it will reach 500 billion by 2025 [16]. It is clearly seen that IIoT will make a remarkable impression on occurring business prototypes in several areas such as energy, agriculture, transportation and many more.

Security issues arise when data is exchanged and when there is an interconnectivity of devices with other devices, servers or platforms. This data is stored by amalgamate cloud servers, and we have to believe these servers [17]. Although these services provide indisputable benefits, these amalgamate IoT systems must also face the possible issues as follows.

10.1.1 INTEROPERABILITY

Interoperability is a process when software and hardware (system) exchange the necessary data or information. It is considered as a substantial challenge in the IIoT systems. Due to insufficient software and hardware, the operation technologies (OTs) and information technologies have to face challenges [18].

10.1.2 DEVICE RELIABILITY AND DURABILITY

It is considered to be very important for devices that are installed in rough industrial environments such as manufacturing, retail, transportation, and so on.

10.1.3 SECURITY AND PRIVACY ISSUES

While using IoT system, the data is not protected and there is risk of data leakage. There must be privacy protection under data protection regulations and the protection of industrial resources, human resources, and evaluative infrastructures while using IoT [19].

10.1.4 EMERGING TECHNOLOGIES AND SKILLS OF STAFF

After introducing new and emerging technology such as IoT at industrial or manufacturing level, it might bring challenging problems to workers. Managing huge data and lack of the management skills bring more problems to labor.

10.1.5 STANDARDIZATION

Lack of standards is considered as the main challenge in blockchain. In the last 20 years, to maintain standardization in the technology, there are continuous efforts being made. While addressing IoT problems and building various networks, compatibility concerns need not be discarded by the device manufacturers [20].

In this chapter, we keep an eye on such challenges faced by industries while implementing blockchain technologies in industry. It is clearly seen that there is enormous change in the digital business of industry when fascinated with blockchain technologies adopted by scholastic and industrial researchers [21]. We will study blockchain in detail and further see the impact of blockchain over the IIoT in Section 10.2 [22].

The rest of the chapter is organized as follows: Section 10.2 discusses the utilization of blockchain technology for Industry 4.0 applications; Industry 4.0 applications based on blockchain technology have been described in Section 10.3; Section 10.4 illustrates open issues in blockchain related to privacy and security, regulations, and resource constraints; Section 10.5 explains the major challenges of implementing blockchain technology into Industry 4.0 followed by the conclusion of the chapter in Section 10.6 [23].

10.2 UTILIZATION OF BLOCKCHAIN FOR INDUSTRY 4.0 APPLICATIONS

There are various factors that rise the demand of blockchain technology for building Industry 4.0 are explained as follows:

10.2.1 DEMAND FOR BLOCKCHAIN TECHNOLOGY IN INDUSTRY 4.0 APPLICATION

Various manufacturing units enjoy many benefits of this blockchain technology, and this technology is considered as a functional tool for Industry 4.0 devices. However, the priority could not be given to blockchain to solve every problem. For the convenience or ease for Industry 4.0 applications, many technologies have been already introduced which performs the same function as blockchain (i.e., creation of a reliable decentralized network that handles the double-spending problem) [24], for example, Direct Acyclic Graph Tangle (DAG) [25]. To resolve the issues of blockchain, the new way of storing data in decentralized and tamper-proof manner is known as Tangle [26]. Tangle is a concede algorithm which is further used by IOTA. IOTA is the new and popular DAG-based project which came into the scene in 2014. Instead of using blocks, Tangle provides high-speed transactions without taking any network fees. Moreover, IOTA could easily collaborate with blockchain, to complete smart agreements. To understand that IOTA and blockchain do not act the same, let us discuss some points: IOTA is different from blockchain, and it is run on non-identical technology known as Tangle (DAG); it does not take any transaction fee and no miners or blocks are involved in this; it is the solution to every problem of IoT [27].

In this way, blockchain is different from IOTA. Before implementing technologies at the industry level or in Industry 4.0 it would be highly recommended to study the features of each technology and recognize them properly. When there is a demand for decentralization in Industry 4.0 technology then the blockchain could be very useful in that case [28]. Not every industry gets benefited from it, but in some industries where a centralized system is not committed, there the demand for this technology is very high. There are numerous service providers, banks, or even various government agencies which are unreliable. If these entities are trustable, then blockchain is not needed [29].

Without faith, transactions could not be performed. With the existing payment system, various systematic tasks automatically start. There are two major drawbacks while adopting transactions through the traditional payment system. Firstly, they charge high transaction fees or they are more expensive than public blockchain [30]. Secondly, without knowing their privacy, security, and morals, we trust them blindly.

P2P communications may be required by many Industry 4.0 applications so that data could be easily transmitted among the non-identical parties. P2P connection is very important and familiar in IIoT architectures. The nodes of this connection cooperate to carry out various jobs.

Every time P2P communications are not the best option for communication and there are so many alternate communication systems that are hidden or not inspected by an Industry 4.0 researcher [31]. Because of more power consumption and more demanding resources, P2P communication could not be easily implemented. Due to this drawback, some IIoT entities or nodes frequently dispatch their information through gateways like AMQP or by making use of edge computing and fog computing. These gateways do not use P2P protocols [32].

Lastly, another challenging problem is the requirement of a proper and vigorous distributed system. Faithful organizations that securely manage the data of critical infrastructures and defense are important in countries especially in those where there is no guarantee of security and privacy [33].

10.2.2   Versatile Nature of Blockchains in Industry

Depending on the requirement of users, there are a wide variety of blockchains. Working of some blockchains is discussed below.

10.2.2.1   Depending on Access Regulation

There are different types of blockchain which are explained as follows:

   i. **Public Blockchains**: Permission of the organization is not required. Everyone has the right to publish and prove transactions [34]. Rewards are also given to miners for their work. In various industries, these public blockchains are considered to be very useful, where the interaction of user devices is necessary. Bitcoin or Ethereum is the best example of this [35].

   ii. **Private Blockchains**: Possessors are responsible for the regulation of blockchain. The owner has the right to grant permission to the user or decides who could control the network and appoints duty to each node according to their validation of the blockchain [36]. Due to the presence of the owner by itself, private blockchains could not be contemplated or decentralized.

   iii. **Federated or Consortium Blockchains**: It is a subpart of private blockchain or said to be a semi-private blockchain [37]. In this, there is the involvement of two or more owners to perform the specific function. The consumer could not directly access the network because the owner is responsible for limiting the user activities. Selected nodes and groups are responsible for the smooth run of the consensus algorithm [38]. Each user has its private valid node, and this node is responsible for consent transaction, and then it is appended to the blockchain technology. IBM food trust is the well-known example of blockchain, where different organizations administered a ledger [39]. We could see consortium blockchain in the energy and power industry and trading or finance sector [40].

10.2.2.2   Depending on Permissions

i. **Permissionless blockchains**: Every individual user is willing to execute respective efforts on the blockchain technology; hence, authority or management consent is not required. The most feasibly accepted permissionless blockchains are Ethereum and Bitcoin.

ii. **Permissioned blockchains**: User could able to perform transactions on the blockchain which are controlled. An example of permission blockchain is Ripple and Monax [41].

10.2.2.3   Depending on the Operation Mode

i. **Logic-oriented blockchains**: It runs on specific logic. It accomplishes certain applications and its most demanding application is smart contracts. Hyperledger-Fabric, Counterparty and NXT are some examples of logic-oriented blockchains [42].

ii. **Transaction-oriented blockchains**: Their objective is to pursue digital strength through logistic operations. Examples are Ripple, Bitcoin, etc.

10.2.2.4   Depending on the Sort of Inducement

i. **Tokenized blockchains**: Their transactions and actions only rely on unique tokens, which are distributed to the users [43]. So, Bitcoin uses Bitcoins and Ethereum uses Ether.

ii. **Non-tokenized blockchains**: They are independent, which means they do not rely on a particular virtual currency. RootStock and Hyperledger-Fabric are such examples of non-tokenized blockchains [44].

10.2.3   Role of Smart Contracts in Industry 4.0 Factories

Blockchain helps many industries to automate various actions, now the time comes to briefly explain the theory of smart contracts. A smart contract is a computer program that accomplishes an agreement and is securely established between at least two groups [45]. This program ensures the performance of various activities when certain conditions satisfy. When all the requirements satisfy then smart contracts are automatically performed [46]. External Services is one that captures the data from actual words and puts it into the blockchain technology, and these facilities are bring up as oracles. This is the condition for smart contracts. For example, to check whether an asset has arrived or not oracles continuously examine data [47]. Block of code executes when a smart contract activates conditional declaration established on the read value [48].

The types of oracles which are based on different types of data gathered from external world are explained as follows:

i. **Hardware Oracles**: From external world, they straight away withdraw information. But there is also a substantial challenge for these hardware oracles, i.e., without data security they describe perusal.

ii. **Software Oracles**: This type of oracles obtains information online. There are various examples that explain this concept; some of these are information of temperature of stored goods, information of secure parts of trucks linked to operational processes [49]. The web is the main source from where data is coming and then it is assembled by validating software. This software acts as an oracle that withdraws the necessary details and plunges them into the smart contract [50].

iii. **Consensus-Based Oracles**: It is a combination of dissimilar oracles to predict the result of occurring. For example, forecast markets like Gnosis and Lowa Electronic Market (LEM) use rating systems for oracles in order to get future results [51].

iv. **Inbound Oracles**: From the outer world, these types of oracles extract information from that sources that do not link with blockchain and then insert them into the blockchain. Comparing to an outbound oracle that permits smart contracts to transmit data to the external world [52].

10.2.4 SUPERIORITY OF ADOPTING BLOCKCHAIN TECHNOLOGY IN INDUSTRY 4.0 TECHNOLOGIES

Due to the involvement of various organizations like operators, suppliers, IIoT nodes, clients, etc. in Industry 4.0 applications, they have to face some common subjects with cryptocurrencies and there is no trust between these organizations [53]. After considering some of the characteristics, these organizations are not similar to cryptocurrencies. For example, take the case of power constraint gadgets like battery-operated machines and IIoT sensors [54]. These devices are directly or indirectly interconnected with the blockchain [55].

When Industry 4.0 resources come into an effective action, they have to face some challenges but with the help of blockchain they successfully handle the four main challenges:

i. In order to run a vertically smart production structure, a network is provided by Industry 4.0 factory. Vertical connection in a smart factory is a connection between two or more institutions that are involved in the value chain of a commodity. After automated connectivity, the gathered information is automatically sent to the applicable elements of the value chain from the numerous systems deployed in a workshop [56]. Vertical integration is easily done by the blockchain, because this technology is furnishing ordinary entrusted data or creating a money interchange place throughout which the numerous factory organizations may interconnect [57].

ii. Constructors, consumers, and suppliers should be managed with horizontally integrated Industry 4.0 technologies [58]. For executing commercial or effortless data transactions, this amalgamation is responsible for necessitating the formation of low-latency and pliable imparting networks, so that organizations involved in Industry 4.0 processes easily execute horizontal integration mechanisms of a blockchain and smart contracts. Additionally, to maintain the communication between users and firms, IIoT appliances and social networks are used as a platform keeping security as a major aspect so that they could also interconnect through a blockchain [59].

iii. Integration is required for Industry 4.0 or smart factories and they amalgamate or integrate throughout the value chain. The main focus of this process is to perform quick actions on the feedback given by the different users in the value chain. That's why some bureaucratic jobs are accelerated by this smart contract and mentioned connections could be performed by blockchains [60].

iv. The integration and advancement of Industry 4.0 with different latest technologies facilitates the working of entrepreneurs and provides a new platform to them on their working domain. The user could easily interchange their information through blockchain and this blockchain acts as intermediate and this is free from all types of technology [61].

10.3   INDUSTRY 4.0 APPLICATIONS BASED ON BLOCKCHAIN TECHNOLOGY

After the introduction of blockchain, many industries get benefited from it and it is used in various applications. But there are still many open issues while using blockchain at various places. Let us take one example to understand how blockchain improves the features of a technology, but the same feature may still be a challenge for another [62]. Cloud computing-based companies may be improved when blockchain is implemented in order to provide redundancy for their storage necessity, but simultaneously due to memory and computational reductions of local blockchain, it is very laborious to duplicate in IIoT entities. It is advisable to look for the welfares and limitations of a blockchain before implementing it in Industry 4.0 appliances [63]. Security regarding aspects such as communication and data availability are increased when blockchain is implemented in Industry 4.0 applications but on the other hand, there are still some obstacles like data privacy, identity certification, etc. The next subsections illustrated the above-disclosed benefits and challenges through blockchain-based IIoT applications [64].

10.3.1   IIoT

When we implement IoT technology at the industry level or at manufacturing units, it will give birth to IIoT or Industry 4.0. A variety of devices such as industrial sensors, machines with remote sensing, actuators are used in IIoT.

Industry 4.0 easily with the help of blockchain is able to perform decentralized transactions and easily exchange information through a committed framework. Nowadays, a credit-based payment scheme is popular whose objective is rapid and periodically deals with the secured energy and it is the best example of blockchain-enabled IoT. Some Industry 4.0 developments based on blockchain technology are blockchain-based IIoT appliances which guarantee that there is no leakage and is not attack by third parties on any type of information published by a particular IIoT device;; While some industries have to give access to their IIoT data to the public or government, stakeholders, and managers in order to make the trust of users; To maintain the proper communication system so that every user could access the data without any interruption, blockchain guaranteed the data access [65]. However, there are still many challenges while implementing blockchain in Industry 4.0 applications:

i. **Data Protection**: Because of hash or public key, each and every IIoT tools can be distinguished. We could achieve high IIoT data protection by mixing techniques, except in different cases in which they can be de-anonymized [66].

ii. **Data Integrity**: IIoT systems are unable to generate trust. Data that is gathered from IIoT devices has not been altered, so in order to overcome this issue, the concept of the third party is introduced. Frameworks having an integrity service are provided by the blockchain in order to neglect malicious parties, but such a sort of organization still has to evolve [67].

iii. **Energy Efficiency**: Many Industry 4.0 applications operate with batteries, so the power utilization should be less for peer-to-peer communications and mining blockchain which acquires a large amount of power. Current IIoT nodes are generally reckoned on in-between gateways that act as an agent with the blockchain technology. They are not allowed to cope with mining or PoW [68].

iv. **Size of Blockchain**: Nowadays, a more number of transactions are performed by IIoT devices, but these gadgets are not able to cope with small blockchains. So, the study about how to compress the blockchains or how to make mini-blockchains should be introduced.

10.3.2 Vertical and Horizontal Integration Systems

Vertical and horizontal integration systems assist suppliers and consumers to communicate with each other inside yards. Basically, this system created such an environment so that data should be exchanged automatically [69]. Manufacturing-Execution System (MES), Enterprise Resource Planning (ERP), IoT platforms and Product Lifecycle Management (PLM) are not able to provide that much kind of integrity because IIoT requires auxiliary integrations levels which are generally very costly [70].

Decentralized transactions and their data administration abilities along with the horizontal integration in manufacturing and supply chain procedures are very useful in order to intensify blockchain [71]. Transaction history supplies transparency to malicious parties and also assists to keep track of the goods [72].

Consumer's supply chain enjoys many benefits from blockchain and these are as follows:

i. Digital recognition for products is created in order to preserve possession and limit simulations.
ii. Tamper-proof records of production of goods, conservation, and handling are provided by it.
iii. Because of smart contracts, the organizations and their objects which are included in the supply chain automatically start tendering [73].
iv. Blockchain is considered an important tool to promote answerable consumption.

10.3.3 ICPS

Cyber-Physical Production System (CPPS) or ICPS is a system that controls the physical processes by collecting data, processing it, and storing it. Smart factories and the Internet enjoy components of ICPS through which they are capable to build real-time commitments with the help of data processing and analysis [74]. Actually, some analysts have already suggested that the blockchain technology is the pillar of the ICPS system for various entities. Many authors offer fair treatments to the various industries that cooperate with the systems in order to increase reliability [75].

10.3.4 ROLE OF BIG DATA AND DATA ANALYTICS

Industry 4.0 gathers an enormous amount of data from distinct entities. Transforming this data requires accelerated big data techniques. Future claims or imminent problems can be concluded by data analytics. Automated reliable data circulation, data collection, and data trustworthiness are the three dominant topics in which big data and data analytics can visage, and blockchain can strengthen both of them while covering these issues. In big data, the data is shared by different entities, and blockchain makes a separate platform via which all these entities can interact. Additionally, by creating positiveness among involved entities and by securing the data, data reliability can be improved by blockchain [76]. Data circulation in Industry 4.0 requires approval by the holder and subject. Smart arrangements can be valuable in this because they make the circulation standardize and imbrute it, without involving any other party.

10.3.5 INDUSTRIAL AUGMENTED AND VIRTUAL REALITY

Virtual elements can be displayed on top of the real-world physical environment by AR device. Environment and the elements both are virtual regardless of virtual reality (VR). Both AR and VR proved that they can act as a helping hand in increasing productivity. They verify to upsurge productivity and can be sympathetic in the industrial design process while synthesizing goods [77]. Not much research has been carried out on blockchain technology over IAR/IVR, even though it is clearly visible that both technologies would gain profit by adopting blockchain in regard to different conditions:

i. Most compatible AR/VR assets are preserved sectionally, but most of the gadgets are restrained with reference to memory and power processing that is why they have to depend upon remote service for storing data. Depending on wearable device requests, these remote servers send AR/VR data. In a similar way, periodic information is sent to the central server by some AR/VR applications which provides functionality and traceability [64].

ii. Data availability can also be improved by blockchain. Large bandwidth is necessary for shuffling of data because when multiple AR/VR devices are connected or convey allochronic with the central server, it might get over-burden and shatter the service.

iii. Enhance data sharing as AR/VR solutions permit the distribution of digital information in the proper way which are proposed by some entities. Additionally, blockchain enhances cybersecurity in data sharing. For example, sharing of important and sensitive data of the military and other forces.

Financial transactions can also be done via AR/VR applications which can be done by using cryptocurrency or tokens and it is based on blockchain. Along with these advantages, some disadvantages are also there, IAR/IVR gadgets based on block-chain technology are battery-operated gadgets which are difficult to design the system [78].

10.3.6 Autonomous Robotics and Vehicles

In Industry 4.0 applications, automation is a very critical field. Therefore, the task impressionable to be computerized can be performed using the following methods.

10.3.6.1 Cobots

After graduation in existing artificial intelligence algorithms, technology is adjusting in order to fulfil our needs. Collaborative robots or industrial cobots are designed in such a way so that they could easily communicate with humans in a guarded and quick mode. Cobots are more efficient than robots because they learn their job more easily than industrial robots.

10.3.6.2 Robots

The Internet of Robotic Things (IoRT) is developed by a combination of robotic groups and IoT. The events happening around us are monitored by intelligent devices which work on the principle of the IoRT. Robots are designed on the concept of machine learning and artificial intelligence, so that they could work in unpredictable conditions.

10.3.6.3 AGV (Autonomous Ground Vehicle)

Transportation of the manufactured product or raw materials could easily do with the autonomous ground vehicle or mobile vehicles. They work on the collaboration of sensor base guidance systems and software. It is less expensive, more flexible, and has more production.

Blockchain can enhance this field by providing a platform by which they can contact other entities via small contracts. Therefore, vehicles and robots based on blockchain technology can collude and perform business with third parties in a sovereign way. Even though there is no much analyses on blockchain carried out over cobots and robots, in recent years, autonomous vehicle exploration displays concerns in the blockchain. Alternative researches are also there which proposed to use the blockchain for refueling, parking, and impeaching. According to the management of robots and vehicles they may be rewarded via a blockchain-based system [79].

10.3.7 CLOUD AND EDGE COMPUTING

Cloud computing system is the trending system on which most modern industrial companies depend because it allows many industry 4.0 Members to conspire among themselves in Easy way. But this system has some detriments also along with advantages, and one of them is that the whole system may get blocked to every user, if the cloud is affected by high call of duty or software problems. Even though cloud system takes care of dissemination of work load and balancing it to avoid such problems, most of them were literally not apprehend from scrape as P2P (peer to peer) system. Blockchain-based system can be sympathetic to resolve this matter as it is designed as an appropriate system that can enhance the cloud computing distribution in assured form [80].

In the last few years, some alternate approaches were proposed to defeat abovementioned obstruction. Examples of this are fog and edge computing. They depend upon offloading part of the processing from cloud to the edge of the network. This also helps to reduce latency response.

10.3.8 ADDITIVE MANUFACTURING (3D PRINTING)

Industry 4.0 paradigm abides resilience and customization. Small factories should provide both these countenance without stepping on the product's cost. Applications of blockchain in industry manufacture are decentralized supply chains, file tracking, increased assurance, intellectual property security and distributed computing.

10.3.9 CYBERSECURITY

It is mandatory to protect these systems which are involved in intra- and interconnection. Both these connections are critical in Industry 4.0 applications. Cybersecurity is also important for the industrial critical systems to avoid cyber-attacks.

The number of attackers can be reduced by private and consortium blockchain. Furthermore, data is distributed in blockchain, one member may be under attack, the data can be accessible using another entity which promises data availability. Therefore, it is important to clear that blockchain is prone to Sybil attacks which are responsible for changing behavior in the system [81].

10.3.10 SIMULATION SOFTWARE

The presence of all the organizations in production systems can be modeled by the information accumulated by the Industry 4.0 factory with the help of simulation software. Examples of entities are operators, products, and machines. This software can conclude the instant state of the factory and then can conclude future events. Even advice precautions to conquer issues or suggest advancement to enhance the quality or to dwindle the cost. The different ways by which simulation software can gain wealth by blockchain technology are gather data from various authorities, authorization of data accuracy which leads to the eviction of uncertainty, blockchain can dispense tasks and estimation among distinct nodes to step up stimulation, and blockchain upgrade stimulation services. It also intensifies decentralized co-stimulation performance.

10.4 OPEN ISSUES IN BLOCKCHAIN TECHNOLOGY

There are various open issues in adopting blockchain technology which are explained as follows.

10.4.1 MARKETABILITY

Marketability controversy in blockchain technology is one of the major shackles in approval and implementation of blockchain. An example of this is VISA. After a successful comparison of VISA and bitcoin, it is concluded that VISA can do more transactions per second than bitcoin. VISA can do near about 2,000 transactions per second whereas bitcoin-only do 7 transactions per second [82]. To tenure the blockchain copy in the current blockchain application, the participating node is mandatory. In comparison to conventional databases, this undergoes to have humongous storage burden. This problem is generally observed in the IIoT scenario generally with increasing sensor nodes and the amount of data generated but the Ethereum like light weighted platform has been developed.

10.4.2 SECURITY AND PRIVACY

Eavesdropping attacks and replay attacks are some of the security attacks through which IoT networks are already prone. Blockchain also provides us with its own set of surveillance indebtedness. Examples of surveillance indebtedness are smart contract program vulnerabilities, message hijack etc. Privacy exposure is another emerging topic in the blockchain.

10.4.3 RESOURCE CONSTRAINTS

Secure network connections are generally required in the utilization of current blockchain. Even though IIoT applications are not always certified with secure network connection, we can say that utilization of blockchain technology is another open challenge due to limitation of resources.

10.4.4 REGULATIONS

Decentralization is one of the properties of blockchain via which transactions can take place between two parties directly without involvement of any intermediary. With technical development, many new blockchain platforms are being recommended, and along with this, the need for enforcement of industry standards for block data formats and interfacing formats has elevated even more.

The popularity of blockchain technology has increased its demand in the following areas as well:

i. **Education Industry**: In today's era, the need and importance of online study increases, and along with this, the need for an autonomous and translucent way of authenticating educational records also increases. Educational institutes and management can have intact transcripts and records via blockchain-based platforms. The increasing population and influential employment path clearly indicate that education or qualification is necessary to have a boon career path. But along with this, qualification duplicity is also increasing. Even though the qualification is crucial not just to apply for a role, but that role can enhance someone's future and this is the reason why qualification duplicity is increasing. However, blockchain can bolster out to clarify this problem and can be pursued as a backbone for educational proof, by determining whether the information given by the candidate is meticulous or not. Massachusetts Institute of Technology (MIT) in the US is implementing this Blockchain technology in their institute by building an app, named block certs wallet through which graduates can share their qualification certificates with staff members [83]. It is uncomplicated to perceive and adopt this technology, just a digital token is required which exhibits qualification, and credentials can be added to the app which grants both a digital degree and immediate affidavit of authenticity.
ii. **Music Industry**: The authenticity of a copyrighted product can be ensured by a smart contract. Accord and synergy between shareholders of products can be delineated by these contracts. They also ensure that the product which customers buy is not just the perfect copy but the authentic product. A potential use case of smart contracts in the music industry is the music right management. In the last 10–15 years, there has been a drastic change in the music industry due to the increase in the Internet and many online music streaming services [84]. Many people have been affected by this like publishers, songwriters, artists, etc. In the music industry, royalty determination mechanisms have always been a complex task but Internet expansion made it even worse. Blockchain technology can be useful in the music industry by making the royalty payment mechanism translucent and this can be done via managing exact and global decentralized databases for accumulating information of music right association in form of the public ledger. Artists are able to create and capture more value from their products as the industry behaves as intermediate. After the royal break up of each work, that work is added to the ledger and the original stakeholders of the copyrighted product are awarded a fair royalty share.

10.5 MAJOR CHALLENGES OF IMPLEMENTING BLOCKCHAIN TECHNOLOGY INTO INDUSTRY 4.0

To every advantage, there is a corresponding disadvantage. Therefore, Industry 4.0 applications have to face various challenges after the implementation of blockchain technology and these challenges require further research:

10.5.1 SCALABILITY

Traffic generated by IIoT applications is usually controlled by the framework which underpin blockchain-based Industry 4.0 applications. Due to mist computing architectures and fog, they are able to supply most primary services adjacent to the point where they are physically necessitated, and for the traditional centralized cloud-based architectures such an amount may arise trouble.

10.5.2 CRYPTOSYSTEMS FOR RESOURCE-CONSTRAINED DEVICES

Many appliances like sensors, tools and actuators in Industry 4.0 have to compete with latest secure public-key cryptography schemes because they have very limited computational resources, in other words, they restrained processing and memory power. Most blockchains based on ECC (i.e., Elliptic Curve Cryptography) and make use of public-key cryptosystems and it is lighter as compare to RSA (i.e., traditional Rivest-Shamir Addleman) keeping security level in mind [85]. Such sort of cryptography consumes more power. Industries which secure data for medium- and long-term should be aware of the possibility of post-quantum computing. Also, these industries look for the algorithms which are riskless and energy-efficient.

10.5.3 CONSENSUS ALGORITHM SELECTION

Selection of the consensus algorithm should be done carefully because it plays a very powerful role in the working of the blockchain. It is advisable not to use the unrealistic and egalitarian concurrence mechanism because while adopting, they give the identical weight to all the miners. These kinds of mechanisms have to suffer from various Sybil attacks and then a single organization takes the command of the entire blockchain. There are many other consensus algorithms besides customary PoW consensus algorithm which is used by Bitcoin, and these are Practical Byzantine Fault Tolerance (PBFT), Proof-of-Personhood or Proof-of-Burn, Proof-of-Stake, and Proof-of-Space. Although mining is convenient in public blockchains, it is not critical for each framework. So, the energy consumed and computational attempts could be drastically decreased which is used by consensus algorithms.

10.5.4 SECURITY AND PRIVACY

Data protection and data integrity are still open concerns that require to be remitted properly. Identity certification is also an open issue. It is important to note when an entity permits a specific identity provider to allow ingress to IIoT gadgets and distinct contributors to the IoT environment, such a provider is also normally able to chunk their approach. Keeping that issue in mind several researchers suggest the use of permissioned blockchains, which are able to control multiple IIoT node identification, while some researchers propose the use of multi chains, in which only few contributors could trace blockchain activities.

10.5.5 ENERGY AND COST EFFICIENCY

We previously discussed the relation of IIoT and CPSs, but they could be extrapolated to other blockchain applications. Factors such as computationally complex cryptographic algorithms, ineffective P2P protocols, and mining have an impression on the energy consumption in every circumstance and when usage of battery-operated gadgets come into role then these factors become condemnatory. Workflow of consensus algorithm and procedure, how blocks are added to the blockchain influenced the above factors.

Each and every single node of Ethereum is responsible for each and every individual transaction inside a network. With the evolution of off-chain solutions such as Plasma and Raiden or on-chain solutions like sharding, Ethereum is making a stable advancement towards scalability. Shards are formed when transactions, and nodes are split into the smaller segments through sharding. Shards have their own transaction record. Consequently, for certain shards, certain nodes would proceed with transactions.

As compared to sharding, Raiden also contributes to scaling the Ethereum network with the help of off-chain transactions. It is said that it is Ethereum's edition of Bitcoin's Lightning Network. Collection of nodes permits to inaugurate bidirectional payment channels in order to ease the fee, near-instant and scalable microtransactions, and in this case, there is zero involvement of Ethereum blockchain. μRaiden is a simplified edition of this solution which was introduced recently. Through unidirectional payment channels, μRaiden is able to formulate micropayments. Plasma is mandatory for the implementation of smart contracts and it is considered as an advanced substructure. Blockchains based on the tree hierarchy concept are created by the plasma. In this, parents are responsible for creation of child blockchains. Plasma Debit, Plasma Cash and Minimal Viable Plasma (MVP) are the three foremost implementations. To tackle with Ethereum's issues there are numerous contrasting substitutes also available. Due to the better centralized character of Electro-Optical System (EOS), it remains controversial but it enhances the scalability and transaction fees of Ethereum [86].

PoW is a consensus algorithm used in the protocol of blockchain. PoW is used during mining blockchain and absorbs a lot of electricity and this is the reason why it is quite lavish. Size of blockchain is directly proportional to time. With increase in time, size of blockchain increases and dynamic prospectors are required to run these algorithms. Some energy-efficient accord protocols which have been intended in modern time are as follows.

10.5.5.1 PoS (Proof-of-Stake)

The cryptocurrency blockchain network attains a strew consensus through consensus type of procedure named as Proof-of-Stake (PoS). The one who created the next block is filtered by the combination of random selection on the basis of wealth or age (i.e., the stake).

10.5.5.2 DPoS (Delighted Proof-of-Stake)

DPoS is also a consensus type of algorithm, which is an advancement or updated version of the Proof-of-Stake (PoS). It is a kind of a system which is managed by an election system for choosing the node named as 'block producers' or 'witnesses.'

10.5.5.3 PoT (Proof-of-Trust)

These protocols do not store the whole blockchain data. Instead of this, they just store the recent transaction data record on blockchain. However, IIOT devices are yet to repulse with the huge amount of industrial data generated. Thus, it is clear that the need of more profitable accord algorithms is crucial and is surely an open challenge.

10.5.6 STORAGE

Blockchain has already shown and even showing its revelation worldwide along with numerous advantages to give a boom to peculiar industries, institutes, banks, etc. And here is another primary benefit of blockchain that is to store transaction and device IDs. It wipes out the demand of the central server, but in this case, ledger should be stored on nodes by themselves. Along with time and the increasing number of nodes, the size of the appropriate ledger will increase. As discussed earlier, IoT devices have low storage capacity and computational revenue.

10.5.7 DEARTH OF ACCOMPLISHMENT

In today's era, blockchain technology is growing worldwide, but still we cannot evade the fact that this technology is contemporary and requires some time to enhance. Even though bank official like, employees employed in banks have a huge amount of intelligence and skills about blockchain, there is a lack of awareness among people about blockchain applications. There are numerous IoT devices worldwide but to adopt blockchain technology in IoT will be difficult without public awareness.

10.5.8 LEGAL AND COMPLAISANCE

Blockchain technology is the technology which can bridge peculiar people from peculiar countries without having any legal or consent code to follow, and for both service providers and manufactures this is a matter of concern. This is an assertion which leads to devise a hurdle for ratifying blockchain in many utilization and businesses.

10.5.9 NAMING AND DISCOVERY

Blockchain technology has not been created for the IoT. That is, nodes were not meant to detect each other in a network. Bitcoin is the example of this, which subsist of IP addresses, performs some operators inside bitcoin applicants and which is used by nodes to build vigorous network topology. As IoT devices keep stunning frequently with time so, it will switch the topology frequently. So, this access will not work out for IoT.

10.5.10 REQUIRED INFRASTRUCTURE

Specific hardware infrastructures like mining hardware or additional storage are expanded by utilization of blockchain technology. Additionally, the huge amount of data generated by the P2P communication system leads to the data traffic, which further needs communication infrastructure and assemblage to support the predicted load.

10.5.11 INTEROPERABILITY AND STANDARDIZATION

We cannot avoid interoperability in companies because it is very useful to gain a flawless integration. Even though in today's era, most of the companies have matured contrasting or their own solutions regarding blockchain, interoperability is crucial among them. Some organizations like IEEE are employed on some explicit standards, which intend assured interoperability in diverse fields. Specifically, some pertinent actions are being driven by the IEEE standard association in relation to blockchain and they are to make accord on advancing clinical preliminary, to increase patient safety, and to drive collusion on promoting blockchain ratification within the pharmaceutical industry.

10.5.12 REGULATORY AND LEGAL ASPECTS

This is another main prerequisite specialized challenge. There are numerous laws and regulations matured by government agencies. So, it is important to reward surveillance to all those laws and regulations. An example of this is that in February 2018, European Union (EU) floated its blockchain forum and observatory, whose main aspirations are to monitor development at the EU level and develop prevalent actions at the EU level.

10.5.13 MANAGEMENT OF MULTICHAINS

Some companies may be required to support several blockchain procreations simultaneously. By using bitcoin, a company may knob its financial transaction, whereas smart contracts are accomplished on utilization that relies on Ethereum. Therefore, to use peculiar blockchains at the same time, there is a need for designed solutions and implementation.

10.6 CONCLUSION

After integrating blockchain technology into the IoT-based industries, it is expected that there would be growth in many industrial sectors and they would gain benefit from it. With the effort of blockchain, one can easily able to integrate trust, security, immutability, decentralization, disintermediation, and a higher degree of automation through smart contracts to Industry 4.0 technologies. After reviewing this chapter, we could easily conclude that blockchain is the solution for the many open challenges in Industry 4.0 technologies. The chapter also studies the main challenges and determines whether the utilization of blockchain is suitable in Industry 4.0 applications or not. This chapter also provides a guide to the future IIoT aspirants and they will able to study the versatile nature of the blockchain. We study all the collected research work of the various researchers on blockchain applicability in numerous Industry 4.0 industries. To unlock all the premium features of blockchain, further industry-intended research is required on the main challenges which include the scalability of the block statistics, data privacy, adoption price, government principles, security of the participating firm, and blockchain integration. Thus, with the use of some latest technologies, Industries 4.0 would able to reshape them or give them a new opportunity.

REFERENCES

1. Puthal, D., Malik, N., Mohanty, S.P., Kougianos, E., & Das, G. (2018). Everything You Wanted to Know About the Blockchain: Its Promise, Components, Processes, and Problems. *IEEE Consumer Electronics Magazine*, 7, 6–14.
2. https://www.industryarc.com/Report/7385/industrial-internet-of-things-(IIoT)-market-report.html.
3. Lin, J., Yu, W., Zhang, N., Yang, X., Zhang, H., & Zhao, W. (2017). A Survey on Internet of Things: Architecture, Enabling Technologies, Security and Privacy, and Applications. *IEEE Internet of Things Journal*, 4, 1125–1142.
4. Huang, J., Kong, L., Chen, G., Wu, M., Liu, X., Zeng, P. (2019). Towards Secure Industrial IoT: Blockchain System with Credit-Based Consensus Mechanism. *IEEE Transactions on Industrial Informatics*, 15, 3680–3689.
5. Goyal, S., Sharma, N., Kaushik, I., Bhushan, B. (2021). Blockchain as a solution for security attacks in named data networking of things. In: *Security and Privacy Issues in IoT Devices and Sensor Networks*, 211–243. doi:10.1016/b978-0-12–821255-4.00010–9.
6. Xu, Y., Ren, J., Wang, G., Zhang, C., Yang, J., Zhang, Y. (2019). A Blockchain-Based Nonrepudiation Network Computing Service Scheme for Industrial IoT. *IEEE Transactions on Industrial Informatics*, 15, 3632–3641.
7. Aggarwal, V.K., Sharma, N., Kaushik, I., Bhushan, B., Himanshu (2021). Integration of Blockchain and IoT (B-IoT): Architecture, Solutions, & Future Research Direction. *IOP Conference Series: Materials Science and Engineering*, 1022, 012103. https://doi.org/10.1088/1757-899x/1022/1/012103.
8. Liang, W., Tang, M., Long, J., Peng, X., Xu, J., Li, K. (2019). A Secure FaBric Blockchain-Based Data Transmission Technique for Industrial Internet-of-Things. *IEEE Transactions on Industrial Informatics*, 15, 3582–3592.
9. Yao, H., Mai, T., Wang, J., Ji, Z., Jiang, C., Qian, Y. (2019). Resource Trading in Blockchain-Based Industrial Internet of Things. *IEEE Transactions on Industrial Informatics*, 15, 3602–3609.
10. Zhao, S., Li, S., Yao, Y. (2019). Blockchain Enabled Industrial Internet of Things Technology. *IEEE Transactions on Computational Social Systems*, 6, 1442–1453.
11. Fernández-Caramés, T.M., Fraga-Lamas, P. (2019). A Review on the Application of Blockchain to the Next Generation of Cybersecure Industry 4.0 Smart Factories. *IEEE Access*, 7, 45201–45218.

12. Shrouf, F., Meré, J.B., Miragliotta, G. (2014). Smart factories in Industry 4.0: A review of the concept and of energy management approached in production based on the Internet of Things paradigm. *2014 IEEE International Conference on Industrial Engineering and Engineering Management*, 697–701.

13. Preuveneers, D., Zudor, E.I. (2017). The Intelligent Industry of the Future: A Survey on Emerging Trends, Research Challenges and Opportunities in Industry 4.0. *Journal of Ambient Intelligence and Smart Environments*, 9, 287–298.

14. Li, Z., Kang, J., Yu, R., Ye, D., Deng, Q., Zhang, Y. (2018). Consortium Blockchain for Secure Energy Trading in Industrial Internet of Things. *IEEE Transactions on Industrial Informatics*, 14, 3690–3700.

15. Kang, J., Yu, R., Huang, X., Maharjan, S., Zhang, Y., Hossain, E. (2017). Enabling Localized Peer-to-Peer Electricity Trading Among Plug-in Hybrid Electric Vehicles Using Consortium Blockchains. *IEEE Transactions on Industrial Informatics*, 13, 3154–3164.

16. Yu, Y., Chen, R., Li, H., Li, Y., Tian, A. (2019). Toward Data Security in Edge Intelligent IIoT. *IEEE Network*, 33, 20–26.

17. Goyal, S., Sharma, N., Kaushik, I., Bhushan, B., Kumar, A. (2020). Blockchain as a Lifesaver of IoT. *Security and Trust Issues in Internet of Things*, 209–237. doi:10.1201/9781003121664-10.

18. Xu, L., He, W., Li, S. (2014). Internet of Things in Industries: A Survey. *IEEE Transactions on Industrial Informatics*, 10, 2233–2243.

19. Varshney, T., Singh, R., Rai, A.K., Sharma, N., Bhushan, B. (2020). Prevailing Privacy and Security Technologies in Smart Cities. *SSRN Electronic Journal*. doi:10.2139/ssrn.3747928.

20. Kiel, D., Arnold, C., Voigt, K. (2017). The Influence of the Industrial Internet of Things on Business Models of Established Manufacturing Companies – A Business Level Perspective. *Technovation*, 68, 4–19.

21. Begam, S., Vimala, J., Selvachandran, G., Ngan, T.T., Sharma, R. (2020). Similarity Measure of Lattice Ordered Multi-Fuzzy Soft Sets Based on Set Theoretic Approach and Its Application in Decision Making. *Mathematics*, 8, 1255.

22. Zelbst, P.J., Green, K.W., Sower, V.E., Bond, P. (2019). The Impact of RFID, IIoT, and Blockchain Technologies on Supply Chain Transparency. *Journal of Manufacturing Technology Management*, *31*(3), 441–457. https://doi.org/10.1108/jmtm-03-2019-0118

23. Vo, T., Sharma, R., Kumar, R., Son, L.H., Pham, B.T., Tien, B.D., Priyadarshini, I., Sarkar, M., Le, T. (2020). Crime rate detection using social media of different crime locations and twitter part-of-speech tagger with brown clustering, 4287–4299.

24. Nguyen, P.T., Ha, D.H., Avand, M., Jaafari, A., Nguyen, H.D., Al-Ansari, N., Van Phong, T., Sharma, R., Kumar, R., Le, H.V., Ho, L.S., Prakash, I., Pham, B.T. (2020). Soft Computing Ensemble Models Based on Logistic Regression for Groundwater Potential Mapping. *Applied Science*, 10, 2469.

25. Sun, Y., Zhang, L., Feng, G., Yang, B., Cao, B., Imran, M. (2019). Blockchain-Enabled Wireless Internet of Things: Performance Analysis and Optimal Communication Node Deployment. *IEEE Internet of Things Journal*, 6, 5791–5802.

26. Nash, P. (2017). Challenges of the industrial Internet of Things. [Online]. Available: https://www.invma.co.uk/blog/iiot-challenges.

27. Bhushan, B., Sharma, N. (2020). Transaction privacy preservations for blockchain technology. *Advances in Intelligent Systems and Computing International Conference on Innovative Computing and Communications*, 377–393. doi:10.1007/978-981-15-5148-2_34.

28. Isaja, M., Soldatos, J. (2018). Distributed Ledger Architecture for Automation, Analytics and Simulation in Industrial Environments. *IFAC-PapersOnLine*, 51, 370–375.

29. Gottheil, A. (2018). Can blockchain address the industrial IOT security? [Online]. Available: http://iiot-world.com/cybersecurity/canblockchain-address-the-industrial-IOT-security.

30. Brass, I., Tanczer, L., Carr, M., Elsden, M., Blackstock, J.J. (2018). Standardising a moving target: The development and evolution of IoT security standards. IoT.

31. Bassi, L. (2017). Industry 4.0: Hope, hype or revolution? *2017 IEEE 3rd International Forum on Research and Technologies for Society and Industry (RTSI)*, 1–6.

32. Zhou, L., Wu, D., Chen, J., & Dong, Z. (2018). When Computation Hugs Intelligence: Content-Aware Data Processing for Industrial IoT. *IEEE Internet of Things Journal*, 5, 1657–1666.

33. Soni, D.K., Sharma, H., Bhushan, B., Sharma, N., Kaushik, I. (2020). Security issues & seclusion in bitcoin system. *2020 IEEE 9th International Conference on Communication Systems and Network Technologies (CSNT)*. doi:10.1109/csnt48778.2020.9115744.

34. Jha, S. et al. (2019). Deep Learning Approach for Software Maintainability Metrics Prediction. *IEEE Access*, 7, 61840–61855.

35. Domova, V., Dagnino, A. (2017). Towards intelligent alarm management in the Age of IIoT. *2017 Global Internet of Things Summit (GIoTS)*, 1–5.

36. Hossain, M.S., Ghulam, M. (2016). Cloud-Assisted Industrial Internet of Things (IIoT) - Enabled Framework for Health Monitoring. *Computer Networks*, 101, 192–202.

37. Sharma, R., Kumar, R., Sharma, D.K., Son, L.H., Priyadarshini, I., Pham, B.T., Bui, D.T., Rai, S. (2019). Inferring Air Pollution from Air Quality Index by Different Geographical Areas: Case Study in India. *Air Quality, Atmosphere, and Health*, 12, 1347–1357.

38. Zhou, K., Liu, T., Zhou, L. (2015). Industry 4.0: Towards future industrial opportunities and challenges. *2015 12th International Conference on Fuzzy Systems and Knowledge Discovery (FSKD)*, 2147–2152.

39. Chen, B., Wan, J., Shu, L., Li, P., Mukherjee, M., Yin, B. (2018). Smart Factory of Industry 4.0: Key Technologies, Application Case, and Challenges. *IEEE Access*, 6, 6505–6519.

40. Sharma, R., Kumar, R., Singh, P.K., Raboaca, M.S., Felseghi, R.-A. (2020). A Systematic Study on the Analysis of the Emission of CO, CO_2 and HC for Four-Wheelers and Its Impact on the Sustainable Ecosystem. *Sustainability*, 12, 6707.

41. Yan, J., Meng, Y., Lu, L., Li, L. (2017). Industrial Big Data in an Industry 4.0 Environment: Challenges, Schemes, and Applications for Predictive Maintenance. *IEEE Access*, 5, 23484–23491.

42. Park, H., Kim, H., Joo, H., Song, J. (2016). Recent Advancements in the Internet-of-Things Related Standards: A oneM2M Perspective. *ICT Express*, 2, 126–129.

43. Jadon, S., Choudhary, A., Saini, H., Dua, U., Sharma, N., Kaushik, I. (2020). Comfy Smart Home Using IoT. *SSRN Electronic Journal*. doi:10.2139/ssrn.3565908.

44. Kim, J., Yun, J., Choi, S., Seed, D., Lu, G., Bauer, M., Al-Hezmi, A., Campowsky, K., Song, J. (2016). Standard-Based IoT Platforms Interworking: Implementation, Experiences, and Lessons Learned. *IEEE Communications Magazine*, 54, 48–54.

45. Varshney, T., Sharma, N., Kaushik, I., Bhushan, B. (2019). Architectural model of security threats & their countermeasures in IoT. *2019 International Conference on Computing, Communication, and Intelligent Systems (ICCCIS)*. doi:10.1109/icccis48478.2019.8974544.

46. Jaidka, H., Sharma, N., Singh, R. (2020). Evolution of IoT to IIoT: Applications & Challenges. *SSRN Electronic Journal*. doi:10.2139/ssrn.3603739.

47. Li, X., Jiang, P., Chen, T., Luo, X., Wen, Q. (2017). A Survey on the Security of Blockchain Systems. *Future Generation Computer Systems* [online] Available: http://www.sciencedirect.com/science/article/pii/S0167739X17318332.

48. Sharma, S. et al. (2020). Global Forecasting Confirmed and Fatal Cases of COVID-19 Outbreak Using Autoregressive Integrated Moving Average Model. *Frontiers in Public Health*. https://doi.org/10.3389/fpubh.2020.580327.

49. Apostolaki, M., Zohar, A., Vanbever, L. (May 2017). Hijacking bitcoin: Routing attacks on cryptocurrencies. *Proceedings of IEEE Symposium on Security and Privacy (SP)*, pp. 375–392.

50. Conti, M, Kumar, E.S., Lal, C., Ruj, S. (2018). A Survey on Security and Privacy Issues of Bitcoin. *IEEE Communications Surveys and Tutorials*, 20(4), 3416–3452, 4th Quart. 2018.

51. Dorri, A., Kanhere, S.S., Jurdak, R. (2019). MOF-BC: A Memory Optimized and Flexible Blockchain for Large Scale Networks. *Future Generation Computer Systems*, 92, 357–373.

52. Brusakova, I.A., Borisov, A.D., Gusko, G.R., Nekrasov, D.Y., Malenkova, K.E. (2017). Prospects for the development of IIOT technology in Russia. *2017 IEEE Conference of Russian Young Researchers in Electrical and Electronic Engineering (EIConRus)*, 1315–1317.

53. Sajid, A., Abbas, H., Saleem, K. (2016). Cloud-Assisted IoT-Based SCADA Systems Security: A Review of the State of the Art and Future Challenges. *IEEE Access*, 4, 1375–1384.

54. Malik, P. et al. (2021). Industrial Internet of Things and Its Applications in Industry 4.0: State-of the Art. *Computer Communication*, 166, 125–139, Elsevier.

55. Varshney, T., Sharma, N., Kaushik, I., Bhushan, B. (2019). Authentication & encryption based security services in blockchain technology. *2019 International Conference on Computing, Communication, and Intelligent Systems (ICCCIS)*. doi:10.1109/icccis48478.2019.8974500.

56. Quarta, D., Pogliani, M., Polino, M., Maggi, F., Zanchettin, A.M., Zanero, S. (2017). An experimental security analysis of an industrial robot controller. *2017 IEEE Symposium on Security and Privacy (SP)*, 268–286.

57. Guan, Z., Lu, X., Wang, N., Wu, J., Du, X., Guizani, M. (2019). Towards Secure and Efficient Energy Trading in IIoT-Enabled Energy Internet: A Blockchain Approach. *Future Generation Computer Systems*, 110, 686–695. https://doi.org/10.1016/j.future.2019.09.027.

58. Analysis of Water Pollution Using Different Physico-Chemical Parameters: A Study of Yamuna River. *Frontiers in Environmental Science*. https://doi.org/10.3389/fenvs.2020.581591.

59. Uhlmann, E., Hohwieler, E., Geisert, C. (2017). Intelligent Production Systems in the Era of Industrie 4.0 – Changing Mindsets and Business Models. *Journal of Machine Engineering*, 17(2), 5–24.

60. Dansana, D. et al. (2021). Using Susceptible-Exposed-Infectious-Recovered Model to Forecast Coronavirus Outbreak. *Computers, Materials & Continua*, 67(2), 1595–1612.

61. Goyal, S., Sharma, N., Bhushan, B., Shankar, A., Sagayam, M. (2021) IoT enabled technology in secured healthcare: Applications, challenges and future directions. In: Hassanien A.E., Khamparia A., Gupta D., Shankar K., Slowik A. (eds) *Cognitive Internet of Medical Things for Smart Healthcare. Studies in Systems, Decision and Control*, vol. 311. Springer, Cham. https://doi.org/10.1007/978-3-030-55833-8_2.

62. Banerjee, M., Lee, J., Choo, K.R. (2017). A Blockchain Future for Internet of Things Security: A Position Paper. *Digital Communications and Networks*, 4, 149–160.

63. Urquhart, L., McAuley, D. (2018). Avoiding the internet of insecure industrial things. *ArXiv*, abs/1801.07207.

64. Manchanda, C., Sharma, N., Rathi, R., Bhushan, B., Grover, M. (2020). Neoteric security and privacy sanctuary technologies in smart cities. *2020 IEEE 9th International Conference on Communication Systems and Network Technologies (CSNT)*. doi:10.1109/csnt48778.2020.9115780.

65. Yiannas, F. (2018). A New Era of Food Transparency Powered by Blockchain. *Innovations: Technology, Governance, Globalization*, 12(1–2), 46–56. doi:10.1162/inov_a_00266.

66. Caro, M.P., Ali, M.S., Vecchio, M., Giaffreda, R. (2018). Blockchain-based traceability in agri-food supply chain management: A practical implementation. *2018 IoT Vertical and Topical Summit on Agriculture - Tuscany (IOT Tuscany)*, May 2018, pp. 1–4.

67. Lucena, P, Binotto, A.P., Momo, F.d.S., Kim, H. (2018). A case study for grain quality assurance tracking based on a blockchain business network, *arXiv preprint arXiv*:1803.07877.

68. Wang, F., Yuan, Y., Zhang, J., Qin, R., Smith, M.H. (2018). Blockchainized Internet of Minds: A New Opportunity for Cyber-Physical-Social Systems. *IEEE Transactions on Computational Social Systems*, 5, 897–906.

69. Vo, M.T., Vo, A.H., Nguyen, T., Sharma, R., Le, T. (2021). Dealing with the Class Imbalance Problem in the Detection of Fake Job Descriptions. *Computers, Materials & Continua*, 68(1), 521–535.

70. Sethi, R., Bhushan, B., Sharma, N., Kumar, R., Kaushik, I. (2020). Applicability of Industrial IoT in Diversified Sectors: Evolution, Applications and Challenges. *Studies in Big Data Multimedia Technologies in the Internet of Things Environment*, 45–67. doi:10.1007/978–981-15-7965-3_4.

71. Müller, J.M., Kiel, D., Voigt, K. (2018). What Drives the Implementation of Industry 4.0? The Role of Opportunities and Challenges in the Context of Sustainability. *Sustainability*, 10, 247.

72. Sachan, S., Sharma, R., Sehgal, A. (2021). Energy Efficient Scheme for Better Connectivity in Sustainable Mobile Wireless Sensor Networks. *Sustainable Computing: Informatics and Systems*, 30, 100504.

73. Esposito, C., Santis, A.D., Tortora, G., Chang, H., Choo, K.R. (2018). Blockchain: A Panacea for Healthcare Cloud-Based Data Security and Privacy? *IEEE Cloud Computing*, 5(1), 31–37. doi:10.1109/mcc.2018.011791712.

74. Rathi, R., Sharma, N., Manchanda, C., Bhushan, B., Grover, M. (2020). Security challenges & controls in cyber physical system. *2020 IEEE 9th International Conference on Communication Systems and Network Technologies (CSNT)*. doi:10.1109/csnt48778.2020.9115778.

75. Ghanem, S. et al. (2021). Lane Detection under Artificial Colored Light in Tunnels and on Highways: An IoT-Based Framework for Smart City Infrastructure. *Complex & Intelligent Systems*. https://doi.org/10.1007/s40747-021-00381-2.

76. Wu, H., Tsai, C. (2018). Toward Blockchains for Health-Care Systems: Applying the Bilinear Pairing Technology to Ensure Privacy Protection and Accuracy in Data Sharing. *IEEE Consumer Electronics Magazine*, 7(4), 65–71. doi:10.1109/mce.2018.2816306.

77. Salahuddin, M.A., Al-Fuqaha, A., Guizani, M., Shuaib, K., Sallabi, F. (2017). Softwarization of Internet of Things Infrastructure for Secure and Smart Healthcare. *Computer*, 50(7), 74–79. doi:10.1109/mc.2017.195.

78. Nugent, T., Upton, D., Cimpoesu, M. (2016). Improving Data Transparency in Clinical Trials Using Blockchain Smart Contracts. *F1000Research*, 5, 2541. doi:10.12688/f1000research.9756.1.

79. Xia, Q., Sifah, E.B., Asamoah, K.O., Gao, J., Du, X., Guizani, M. (2017). MeDShare: Trust-Less Medical Data Sharing Among Cloud Service Providers via Blockchain. *IEEE Access*, 5, 14757–14767. doi:10.1109/access.2017.2730843.

80. Rustagi, A., Manchanda, C., Sharma, N. (2020). IoE: A boon & threat to the mankind. *2020 IEEE 9th International Conference on Communication Systems and Network Technologies (CSNT)*. doi:10.1109/csnt48778.2020.9115748.

81. Nizamuddin, N, Hasan, H., Salah, K., Iqbal, R. (2019). Blockchain-Based Framework for Protecting Author Royalty of Digital Assets. *Arabian Journal for Science and Engineering*, 44(4), 3849–3866 [online] Available: https://link.springer.com/article/10.1007/s13369-018-03715-4#citeas.

82. Yuan, Y., Wang, F.-Y. (2016). Towards blockchain-based intelligent transportation systems. *Proceedings on IEEE 19th International Conference on Intelligent Transportation Systems (ITSC)*, pp. 2663–2668.

83. Rossow, A. (2019). Hailing rides down crypto lane: The future of ridesharing, [online] Available: https://www.forbes.com/sites/andrewrossow/2018/07/18/hailing-rides-down-crypto-lane-the-future-of-ridesharing/.

84. Holland, M., Stjepandić, J., Nigischer, C. (2018). Intellectual property protection of 3D print supply chain with blockchain technology. *Proceedings on IEEE International Conference on Engineering, Technology and Innovation*, 1–8.

85. Suárez-Albela, M., Fraga-Lamas, P., Fernández-Caramés, T.M. (2018). A Practical Evaluation on RSA and ECC-Based Cipher Suites for IoT High-Security Energy-Efficient Fog and Mist Computing Devices. *Sensors*, 18(11), 3868.

86. Yang, Z., Zheng, K., Yang, K., Leung, V.C.M. (2017). A blockchain-based reputation system for data credibility assessment in vehicular networks. *Proceedings on IEEE International Symposium on Personal, Indoor and Mobile Radio Communications (PIMRC)*, 1–5.

Index

Printed in the United States
by Baker & Taylor Publisher Services

Printed in the United States
by Baker & Taylor Publisher Services